HANDS-ON
MATRIX ALGEBRA
USING R

Active and Motivated Learning with Applications

HANDS-ON
MATRIX ALGEBRA USING R

Active and Motivated Learning with Applications

Hrishikesh D Vinod

Fordham University, USA

World Scientific

NEW JERSEY · LONDON · SINGAPORE · BEIJING · SHANGHAI · HONG KONG · TAIPEI · CHENNAI

Published by

World Scientific Publishing Co. Pte. Ltd.

5 Toh Tuck Link, Singapore 596224

USA office: 27 Warren Street, Suite 401-402, Hackensack, NJ 07601

UK office: 57 Shelton Street, Covent Garden, London WC2H 9HE

British Library Cataloguing-in-Publication Data
A catalogue record for this book is available from the British Library.

HANDS-ON MATRIX ALGEBRA USING R
Active and Motivated Learning with Applications

ISBN-13 978-981-4313-68-1
ISBN-10 981-4313-68-8
ISBN-13 978-981-4313-69-8 (pbk)
ISBN-10 981-4313-69-6 (pbk)

Printed in Singapore.

To my wife Arundhati, daughter Rita and her children Devin and Troy

Preface

In high school, I used to like geometry better than algebra or arithmetic. I became excited about matrix algebra after my teacher at Harvard, Professor Wassily Leontief, Nobel laureate in Economics showed me how his input-output analysis depends on matrix inversion. Of course, inverting a 25×25 matrix was a huge deal at that time. It got me interested in computer software for matrix algebra tasks. This book brings together my two fascinations, matrix algebra and computer software to make the algebraic results fun to use, without the drudgery of patient arithmetic manipulations.

I was able to find a flaw in Nobel Laureate Paul Samuelson's published work by pointing out that one of his claims for matrices does not hold for scalars. Further excitement came when I realized that Italian economist Sraffa's work, extolled in Professor Samuelson's lectures can be understood better in terms of eigenvectors. My interest about matrix algebra further increased when I started working at Bell Labs and talking to many engineers and scientists. My enthusiasm for matrix algebra increased when I worked with my friend Sharad Sathe on our joint paper Sathe and Vinod (1974). My early publication in Econometrica on joint production, Vinod (1968), heavily used matrix theory. My generalization of the Durbin-Watson test in Vinod (1973) exploited the Kronecker product of matrices. In other words, a study of matrix algebra has strongly helped my research agenda over the years.

Research oriented readers will find that matrix theory is full of useful results, ripe for applications in various fields. The hands-on approach here using the R software and graphics hopes to facilitate the understanding of results, making such applications easy to accomplish. An aim of this book is to facilitate and encourage such applications.

The primary motivation for writing this book has been to make learning of matrix algebra fun by using modern computing tools in R. I am assuming that the reader has very little knowledge of R and am providing some help with learning R. However, teaching R is not the main purpose, since on-line free manuals are available. I am providing some tips and hints which may be missed by some users of R. For something to be fun, there needs to be a reward at the end of an effort. There are many matrix algebra books for those purists who think learning matrix algebra is a reward in itself. We take a broader view of a researcher who wants to learn matrix algebra as a tool for various applications in sciences and engineering. Matrices are important in statistical data analysis. An important reference for Statistics using matrix algebra is Rao (1973).

This book should appeal to the new generation of students, "wired differently" with digitally nimble hands, willing to try difficult concepts, but less skilled with arithmetic manipulations. I believe this generation may not have a great deal of patience with long tedious manipulations. This book shows how they can readily create matrices of any size and satisfying any properties in R with random entries and then check if any alleged matrix theory result is plausible. A fun example of Fibonacci numbers is used in Sec. 17.1.3 to illustrate inaccuracies in floating point arithmetic of computers. It should be appealing to the new generation, since many natural (biological) phenomena follow the pattern of these numbers, as they can readily check on Google.

This book caters to students and researchers who do not wish to emphasize proofs of algebraic theorems. Applied people often want to 'see' what a theorem does and what it might mean in the context of several examples, with a view to applying the theorem as a practical tool for simplifying or deeply understanding some data, or for solving some optimization or estimation problem.

For example, consider the familiar regression model

$$y = X\beta + \epsilon, \tag{0.1}$$

in matrix notation, where y is a $T \times 1$ vector, X is $T \times p$ matrix, β is a $p \times 1$ vector and ε is $T \times 1$ vector. In statistics it is well known that $b = (X'X)^{-1}X'y$ is the ordinary least squares (OLS) estimator minimizing error sum of squares $\varepsilon'\varepsilon$.

It can be shown using some mathematical theorems that a deeper understanding of the X matrix of regressors in (0.1) is available provided one computes a 'singular value decomposition' (SVD) of the X matrix. The theorems show that when a 'singular value' is close to zero, the matrix of regressors is 'ill-conditioned' and regression computations and statistical inference based on computed estimates are often unreliable. See Vinod (2008a, Sec. 1.9) for econometric examples and details.

The book does not shy away from mentioning applications making purely matrix algebraic concepts like the SVD alive. I hope to provide a motivation for learning them as in Chapter 16. Section 16.8 in the same Chapter uses matrix algebra and R software to expose flaws in the popular Hodrick-Prescott filter, commonly used for smoothing macroeconomic time series to focus on underlying business cycles. Since the flaw cannot be 'seen' without the matrix algebra used by Phillips (2010) and implemented in R, it should provide further motivation for learning both matrix algebra and R. Even pure mathematicians are thrilled when their results come alive in R implementations and find interesting applications in different applied scientific fields.

Now I include some comments on the link between matrix algebra and computer software. We want to use matrix algebra as a tool for a study of some information and data. The available information can be seen in any number of forms. These days a familiar form in which the information might appear is as a part of an 'EXCEL' workbook popular with practitioners who generally need to deal with mixtures of numerical and character values including names, dates, classification categories, alphanumeric codes, etc. Unfortunately EXCEL is good as a starting point, but lacks the power of R.

Matrix algebra is a branch of mathematics and cannot allow fuzzy thinking involving mixed content. Its theorems cannot apply to mixed objects without important qualifications. Traditional matrices usually deal with purely numerical content. In R traditional algebraic matrices are objects called 'matrix,' which are clearly distinguished from similar mixed objects needed by data analysts called 'data frames.' Certain algebraic operations on rows and columns can also make sense for data frames, while not others. For example, the 'summary' function summarizes the nature of information in a column of data and is a very fundamental tool in R.

EXCEL workbooks can be directly read into R as data frame objects after some adjustments. For example one needs to disallow spaces and certain symbols in column headings if a workbook is to become a data frame object. Once in R as a data frame object, the entire power of R is at our disposal including superior plotting and deep numerical analysis with fast, reliable and powerful algorithms. For a simple example, the reader can initially learn what a particular column has, by using the 'summary' on the data frame.

This book will also review results related to matrix algebra which are relevant for numerical analysts. For example, inverting ill-conditioned sparse matrices and error propagation. We cover several advanced topics believed to be relevant for practical applications. I have attempted to be as comprehensive as possible, with a focus on potentially useful results. I thank following colleagues and students for reading and suggesting improvements to some chapters of an earlier version: Shapoor Vali, James Santangelo, Ahmad Abu-Hammour, Rossen Trendafilov and Michael Gallagher. Any remaining errors are my sole responsibility.

H. D. Vinod

Contents

Chapter 1

R Preliminaries

1.1 Matrix Defined, Deeper Understanding Using Software

Scientists and accountants often use a set of numbers arranged in rectangular arrays. Scientists and mathematicians call rectangular arrays matrices. For example, a 2×3 matrix of six numbers is defined as:

$$A = \{a_{ij}\} = \begin{bmatrix} a_{11} & a_{12} & a_{13} \\ a_{21} & a_{22} & a_{23} \end{bmatrix} = \begin{bmatrix} 3 & 5 & 11 \\ 6 & 9 & 12 \end{bmatrix}, \tag{1.1}$$

where the first subscript $i = 1, 2$ of a_{ij} refers to the row number and the second subscript $j = 1, 2, 3$ refers to the column number.

Such matrices and their generalizations will be studied in this book in considerable detail. Rectangular arrays of numbers are called spreadsheets or workbooks in accounting parlance and 'Excel' software has standardized them in recent decades. The view of rectangular arrays of numbers by different professions can be unified. The aim of this book is to provide tools for developing a deeper understanding of the reality behind rectangular arrays of numbers by applying many powerful results of matrix algebra along with certain software and graphics tools.

Exercise 1.1.1: Construct a 2×2 matrix with elements $a_{ij} = i + j$. [Hint: the first row will have numbers 2,3 and second row will have 3,4].

Exercise 1.1.2: Given integers 1 to 9 construct a 3×3 matrix with first row having numbers 1 to 3, second row with numbers 4 to 6 and last row with numbers 7 to 9.

The R software allows a unified treatment of all rectangular arrays as 'data frame' objects. Although matrix algebra theorems do not directly apply to data frames, they are a useful preliminary, allowing matrix algebra to be a very practical and useful tool in everyday life, not some esoteric subject for scientists and mathematicians. For one thing, the data frame

1

objects in R software allow us to name the rows and columns of matrices in a meaningful way. We can, of course, strip away the row-column names when treating them as matrices. Matrix algebra is abstract in the sense that its theorems hold true irrespective of row-column names. However, the deeper understanding of the reality behind those rectangular arrays of numbers requires us to have easy access to those names when we want to interpret the meaning of matrix algebra based conclusions.

1.2   Introduction, Why R?

The New York Times, 6 January 2009 had an article by Daryl Pregibon, a research scientist at *Google* entitled "Data Analysts Captivated by R's Power." It said that the software and graphics system called R is fast becoming the *lingua franca* of a growing number of data analysts inside corporations and academia. Countless statisticians, engineers and scientists without computer programming skills find R "easy to use." It is hard to believe that R is free. The article also stated that "R is really important to the point that it's hard to overvalue it."

R is numerically one of the most accurate languages, perhaps because it is free with transparent code which can be checked for accuracy by almost anyone, anywhere. R is supported by volunteer experts from around the world and available anywhere one has access to (high speed) Internet. R works equally well on Windows, Linux, and Mac OS X computers. R is based on an earlier public domain language called S developed in Bell Labs in 1970's. I had personally used S when I was employed at Bell Laboratories. At that time it was available only on UNIX operating system computers based on principles of object oriented languages. S-PLUS is the commercial version of S, whereas R is the free version. It is a very flexible object oriented language (OOL) in the sense that inputs, data and outputs are all objects (e.g., files) inside the computer.

R is powerful and fun: R may seem daunting at first to someone who has never thought of himself or herself as a programmer. Just as anyone can use a calculator without being a programmer, anyone can start using R as a calculator. For example, when balancing a checkbook R can help by adding a bunch of numbers '300+25.25+33.26+12*6', where the repeated number 12 need not be typed six times into R. The advantage with R over a calculator is that one can see the numbers being added and correct any errors on the go. The tip numbered xxvi among the tips listed in Sec. 1.5

shows how to get the birth day-of-the-week from a birthday of a friend. It can also give you the date and day-of-the-week 100 days before today and fun things like that. There are many data sets already loaded in R. The dataset 'women' contains Average Heights and Weights for American Women. The simple R command 'summary(women)' gives basic descriptive statistics for the data (minimum, maximum, quartiles, median and mean). No calculator can do this so conveniently. The dataset named 'co2' has Mauna Loa Atmospheric CO2 Concentration to check global warming. The reader will soon discover that it is more fun to use R rather than any calculator.

Actually, easy and powerful graphics capabilities of R make it fun for many of my students. The command 'example(points)' gives code examples showing all kinds of lines along points and creation of sophisticated symbols and shapes in R graphs. A great Internet site for all kinds of R graphics will convince anyone how much fun one can have with R: `http://AddictedToR.free.fr/graphiques/index.php`

The reason why R a very powerful and useful calculator is that thousands of 'functions' and program packages are already written in R. The packages are well documented in standard format and have illustrative examples and user-friendly vignettes. The user simply has to know what the functions do and go ahead and use them *at will*, for free. Even if R is a "language" it is an "interpreted" language similar to a calculator, and the language C, not "compiled" language similar to FORTRAN, GAUSS, or similar older languages.

Similar to a calculator, all R commands are implemented as they are received by R (typed). Instead of having subroutines similar to FORTRAN, R has 'functions.' Calculations requiring hundreds of steps can be defined as 'functions' and subsequently implemented by providing values for the suitable number of arguments to these functions. A typical R package has several functions and data sets. R is functional and interpreted computer language which can readily import and employ the code for functions written in other languages including C, C++, FORTRAN, among others.

John M. Chambers, one of my colleagues at Bell Labs, is credited with creating S and helping in the creation of R. He has published an article entitled "The Future of R" in the first issue (May 2009) of the 'R Journal' available on the Internet at:

`http://journal.r-project.org/\break2009-1/RJournal\`
`_2009-1\_Chambers.pdf`

John lists six facets of R

 (i) an interface to computational procedures of many kinds;

 (ii) interactive, hands-on in real time;

(iii) functional in its model of programming;

(iv) object-oriented, "everything is an object";

 (v) modular, built from standardized pieces; and,

(vi) collaborative, a world-wide, open-source effort.

This list beautifully explains the enormous growth and power of R in recent years. Chambers goes on to explain how R functions are themselves objects with their own functionality.

I ask my students to first type a series of commands into a text editor and then copy and paste them into R. I recommend the text editor called Tinn-R, freely available at: (`http://www.sciviews.org/Tinn-R/`). The reader should go to the bottom of the web page and download it to his or her computer. Next, click on "Setup for Tinn-R" Be sure to use the old stable version (1.17.2.4) (.exe, 5.2 Mb) compatible with Rgui in SDI or MDI mode. Tinn-R color codes the R commands and provides helpful hints regarding R syntax in the left column entitled 'R-card.' Microsoft Word is also a good text editor to use, but care is needed to avoid smart or slanted quotation marks (unknown to R).

1.3 Obtaining R

One can *Google* the word r-project and get the correct Internet address. The entire software is 'mirrored' or repeated at several sites on the Internet around the world. One can go to

(`http://www.r-project.org`) and choose a geographically nearby mirror. For example, US users can choose

`http://cran.us.r-project.org`

Once arriving at the appropriate Internet site, one looks on the left hand side under "Download." Next, click on CRAN Link. Sometimes the site asks you to pick a mirror closest to your location again. Click on 'Windows' (if you have a Windows computer) and then Click on 'base.'

For example, if you are using Windows XP computer and if the latest version of R is 2.9.1 then click on "Download R-2.9.1 for Windows". It is a good idea to download the setup program file to a temporary directory which usually takes about 7-8 minutes if you have a reasonably fast connection to the Internet. Now, double click on the 'setup' file (R-2.9.1-win32.exe) to get R set up on any local computer. When the setup wizard will first ask you language, choose English. Then click 'Next' and follow other on-screen instructions.

If you have a Windows Vista computer, among frequently asked question, they have answer to the question: How do I install R when using Windows Vista? One of the hints is 'Run R with Administrator privileges in sessions where you want to install packages.'

Choose simple options. Do not customize it. The setup does everything, including creating an Icon for R-Gui (graphical user interface). It immediately lets you know what version is being used.

1.4   Reference Manuals in R

Starting at the R website (`http://cran.r-project.org`) left column click on 'manuals' and access the first bullet point called "An Introduction to R," which is the basic R manual with about 100 pages. It can be browsed in html format at:

`http://cran.r-project.org/doc/manuals/R-intro.html`

It can also be downloaded in the pdf format by clicking at a link within that bullet as:

`http://cran.r-project.org/doc/manuals/R-intro.pdf`

The reader is encouraged to read all chapters, especially Chapter 5 of 'R-intro' dealing with arrays and Chapter 12 dealing with graphics.

There are dozens of free books available on the Internet about R at:

`http://cran.r-project.org/doc/contrib`

The books are classified into 12 books having greater than 100 pages starting with "Using R for Data Analysis and Graphics - Introduction, Examples and Commentary" by John Maindonald. Section 7.8 of the above pdf file deals specifically with matrices and arrays in R.

`http://cran.r-project.org/doc/contrib/usingR.pdf`

There are additional links to some 18 (as of July 2009) short books (having less than 100 pages). The list starts with "R for Beginners" by Emmanuel Paradis.

Social scientists and economists have found my own book Vinod (2008a) containing several examples of R code for various tasks a useful reference. The website of the book also contains several solved problems among its exercises. A website of a June 2009 conference on R in social sciences at Fordham University in New York is: `http://www.cis.fordham.edu/QR2009`. The reader can view various presentations by distinguished researchers who use R.

1.5 Basic R Language Tips

Of course, the direct method of learning any language is to start using it. I recommend learning the assignment symbols '=' or '<-' and the combine symbol 'c' and start using R as a calculator. A beginning user of R should consult one or more of the freely available R manuals mentioned in Sec. 1.4. Nevertheless, this section lists some practical points useful to beginning users, as well as, some experienced users of R.

(i) Unlike paper and pencil mathematics, (*) means multiply; either the hat symbol ($\wedge$) or (**) are used for raising to a power; (|) means the logical 'or' and (!) means logical negation.

(ii) If a command is inside simple parentheses (not brackets) it is printed to the screen.

(iii) Colon (:) means generate a sequence. For example, x=3:7 creates an R (list or vector) object named 'x' containing the numbers (3,4,5,6,7).

(iv) Generally speaking, R ignores all spaces inside command lines. For example, 'x=1:3' is the same as 'x = 1 : 3'.

(v) Comments within R code are put anywhere, starting with a hash-mark ('#'), such that everything from that point to the end of the line is a comment by the author of the code (usually to oneself) and completely ignored by the R processor.

(vi) 'c' is a very important and very basic function in R which combines its arguments. One cannot really use R without learning to use the 'c' function. For example, 'x=c(1:5, 17, 99)' will create a 'vector' or a 'list' object called 'x' with elements: (1, 2, 3, 4, 5, 17, 99). The 'c' stands for 'concatenate', 'combine' or 'catalog.'

(vii) Generally, each command should be on a separate line of code. If one wants to type or have two or more commands on one line, a semi-colon (;) must be used to separate them.

(viii) R is **case sensitive** similar to UNIX. For example, the lower case named object b is different from upper case named object B.

(ix) The object names usually start with a letter, never with numbers. Special characters including underscores are not allowed in object names but periods (.) are allowed.

(x) R may be used as a calculator. The usual arithmetic operators +, -,

*, / and the hat symbol ($\wedge$) for raising to a power are available. R also has log, exp, sin, cos, tan, sqrt, min, max with self-explanatory meanings.

(xi) Expressions and assignments are distinct in R. For example, if you type the expression:

2+3+4^2

R will proceed to evaluate it. If you assign the expression a name 'x' by typing:

x=2+3+4^2; x

The resulting object called 'x' will contain the evaluation of the expression and will be available under the name 'x,' but not automatically printed to the screen. Assuming that one wants to see R print it, one needs to type 'x' or 'print(x)' as a separate command. We have included 'x' after the semi-colon (;) to suggest a new command asking R to print the result to the screen.

(xii) The R language and its packages together have tens of thousands of functions for doing certain tasks on their arguments. For example, 'sqrt' is a function in R which computes the square root of its argument. It is important to use simple parentheses as in 'print(x)' or 'sqrt(x)' when specifying the arguments of R functions. Using curly braces or brackets instead of parentheses will give syntax error (or worse confusion) in R.

(xiii) Brackets are used to extract elements of an array. For example, 'x[2]' extracts second element of a list 'x'. Similarly, 'x[2,1]' extracts the number along the second row and first column of a matrix x. Again, using curly braces or parentheses will not work for this purpose.

(xiv) Curly braces ('{' and '}') are used to combine several expressions into one procedure.

(xv) Re-executing previous commands is done by using vertical arrow keys on the keyboard. Modifications to such commands can be made by using horizontal arrows on the keyboard.

(xvi) Typing 'history()' provides a list of all recent commands typed by the user. These commands can then be copied into MS-Word or any text editor, suitably changed, copied and pasted back into R.

(xvii) R provides numerical summaries of data. For example, let us use the vector of seven numerical values mentioned above as 'x' and let us issue certain commands in R (especially the summary command) collected in the following R software input snippet.

```
#R.snippet
x=c(1:5, 17, 99)
summary(x)
```

Then R will evaluate them and report to the screen the following output.

```
   Min. 1st Qu.  Median   Mean 3rd Qu.    Max.
   1.00    2.50    4.00  18.71   11.00   99.00
```

Note that the function 'summary' computes the minimum, first quartile (25% of data are below this and 75% are above this), median, third quartile (75% of data are below this and 25% are above this), and the maximum.

(xviii) Attributes (function 'attr') can be used to create a matrix. The following section discusses the notion of 'attributes' in greater detail. It is particularly useful for matrix algebra.

```
#R.snippet
x=c(1:4, 17, 99) # this x has 6 elements
attr(x,"dim")=c(3,2) #this makes x a 3 by 2 matrix
x
```

Then R will place the elements in the object x column-wise into a 3 by 2 matrix and report:

```
     [,1] [,2]
[1,]    1    4
[2,]    2   17
[3,]    3   99
```

(xix) R has extensive graphics capabilities. The plot command is very powerful and will be illustrated at several places in this book.

(xx) R can by tailor-made for specific analyses.

(xxi) R is an Interactive Programming language, but a set of R programs can be submitted in a batch mode also.

(xxii) R distinguishes between the following 'classes' of objects: "numeric", "logical", "character." "list", "matrix", "array", "factor" and "data.frame." The 'summary' function is clever enough to do the right summarizing upon taking into account the class.

(xxiii) It is possible to convert objects from one 'class' to another within limits. Let us illustrate this idea. Given a vector of numbers, we can convert into a logical vector containing only two items TRUE and

FALSE. by using the function 'as.logical'. All numbers (positive or negative) are made TRUE and the number zero is made FALSE.
Conversely, given a logical vector, we can convert it into numerical vector by using the function 'as.numeric.' The reader should see that this converts TRUE to the number 1 and FALSE to the number 0.
The R function 'as.complex' converts the numbers into complex numbers with the appropriate coefficient (=0) for the imaginary part of the complex number identified by the symbol $i = \sqrt{(-1)}$. The 'as.character' function places quotes around the elements of a vector.

```
# R program snippet 1.5.1 (as.something) is next.
x=c(0:3, 17, 99) # this x has 6 elements
y=as.logical(x);y#makes 0=FALSE all numbers=TRUE
as.numeric(y)#evaluate as 1 or 0
as.complex(x)#will insert i
as.character(x) #will insert quotes
```

The following illustrates how R converts between classes using the 'as.something' command. The opeartion of 'as.matrix' and 'as.array' will be illustrated in the following Sec. 1.7.

```
> y=as.logical(x);y#
[1] FALSE  TRUE   TRUE   TRUE   TRUE   TRUE
> x=c(0:3, 17, 99) # this x has 6 elements
> y=as.logical(x);y#makes 0=FALSE all numbers=TRUE
[1] FALSE  TRUE   TRUE   TRUE   TRUE   TRUE
> as.numeric(y)#evaluate as 1 or 0
[1] 0 1 1 1 1 1
> as.complex(x)#will insert i
[1]   0+0i  1+0i  2+0i  3+0i 17+0i 99+0i
> as.character(x) #will insert quotes
[1] "0"   "1"   "2"   "3"   "17" "99"
```

(xxiv) Missing data in R are denoted by "NA" without the quotation marks. These are handled correctly by R. In the following code we deliberately make the x value along third row and second column as NA, that is we specify that x[3,2] as missing. The exclamation symbol (!) is used as a logical 'not'. Hence if 'is.na' means is missing, '!is.na' means 'is not missing'. The bracketed conditions in R with repetition of 'x' outside and inside the brackets are very useful for subset selection. For example, x[!is.na(x)] and x[is.na(x)] in the following snippet create subsets to x containing only the non-missing values and only the missing values, respectively.

```
# R program snippet 1.5.2 is next.
x=c(1:4, 17, 99) # this x has 6 elements
attr(x,"dim")=c(3,2) #this makes x a 3 by 2 matrix
x[3,2]=NA
y=x[!is.na(x)] #picks only non-missing subset of x
z=x[is.na(x)] #picks only the missing subset of x
summary(x);summary(y);summary(z)
```

Then R will recognize that 'x' is a 3 by 2 matrix and the 'summary' function in the snippet 1.5.2 will compute the summary statistics for the two columns separately. Note that 'y' converts 'x' matrix into an array of 'non-missing' set of values and computes their summary statistics. The vector 'z' contains only the one missing value in the location at row 3 and column 2. Its summary (under column for second variable 'V2' below correctly recognizes that one item is missing.

```
> summary(x);
        V1                V2
 Min.   :1.0    Min.    : 4.00
 1st Qu.:1.5    1st Qu.: 7.25
 Median :2.0    Median :10.50
 Mean   :2.0    Mean   :10.50
 3rd Qu.:2.5    3rd Qu.:13.75
 Max.   :3.0    Max.   :17.00
                NA's    : 1.00

> summary(y)
   Min. 1st Qu.  Median   Mean 3rd Qu.    Max.
   1.0     2.0     3.0    5.4     4.0    17.0

> summary(z)
   Min. 1st Qu.  Median   Mean 3rd Qu.    Max.    NA's
                                                    1
```

Recall that 'y' picks out only the non-missing values, whereas z picks out the one missing value denoted by 'NA'. Note also that R does not foolishly try to find the descriptive statistics of the missing value.

(xxv) R admits character vectors. For example,

```
#R.snippet
x=c("In", "God", "We",  "Trust", ".")
x; print(x, quote=FALSE)
```

We need the print command to have the options 'quote=FALSE' to prevent printing of all individual quotation marks as in the output below.

```
[1] "In"    "God"    "We"    "Trust" "."
[1] In    God    We    Trust .
```

Character vectors are distinguished by quotation marks in R. One can combine character vectors with numbers without gaps by using the 'paste' function of R and the argument separation with a blank as in the following snippet.

```
#R.snippet
paste(x,1:4, sep="")
paste(x,1:4, sep=".")
```

Note that 'x' contains five elements including the period (.). If we attach the sequence of only 4 numbers, R is smart and recycles the numbers 1:4 again and again till needed. The argument '(sep="")' is designed to remove any space between them. If sep=".", R wil place the (.) in between. Almost anything can be placed between the characters by using the 'sep' option.

```
[1] "In1"    "God2"    "We3"    "Trust4" ".1"
> paste(x,1:4, sep=".")
[1] "In.1"    "God.2"    "We.3"    "Trust.4" "..1"
```

(xxvi) Date objects. Current date including time is obtained by the command 'date(),' where the empty parentheses are needed.

```
date()
[1] "Sat Jul 11 14:24:38 2009"
Sys.Date()
weekdays(Sys.Date()-100) #plural weekday
weekdays(as.Date("1971-09-03"))
#find day-of-week from birthdate
```

The output of the above commands is given below. Note that only the system date (without the time) is obtained by the command 'Sys.Date()'. It is given in the international format as YYYY-mm-dd for year, month and day.

Find the date 30 days from the current date:

If you want to have the date after 30 days use the command
'Sys.Date()+30'.

Find the day of week 100 days before the current date:
If you want to know the day of week 100 days before today, use the R
command 'weekdays(Sys.Date()-100).' If you want to know the day of
the week from a birth-date, which is on September 3, 1971, then use
the command 'weekdays(as.Date("1971-09-03"))' with quotes as shown,
and R will respond with the correct day of the week: Friday.

```
> date()
[1] "Sat Jul 11 20:04:41 2009"
> Sys.Date()
[1] "2009-07-11"
> weekdays(Sys.Date()-100) #plural weekday
[1] "Thursday"
> weekdays(as.Date("1971-09-03"))
#find day-of-week from birth-date
[1] "Friday"
```

These are some of the fun uses of R. An R package called 'chron',
James and Hornik (2010), allows further manipulation of all kinds of
date objects.

(xxvii) Complex number $i = \sqrt{(-1)}$ is denoted by '1i'. For example, if x=1+2i
and y=1-2i, the product $xy = 1^2 + 2^2 = 5$ as seen in the following
snippet

```
# R program snippet 1.5.3 (dot product) is next.
x=1+2*1i  #number 1 followed by letter i is imaginary i in R
y=1-2*1i
x*y #dot product of two imaginary numbers

> x*y #dot product of two imaginary numbers
[1] 5+0i
```

1.6   Packages within R

Recall from Sec. 1.3 that John Chambers has noted that R is modular
and collaborative. The 'modules' mostly contain so-called 'functions' in R,
which are blocks of R code which follow some rules of syntax. Blocks of
input code submitted to R are called snippets in this book. These can call
existing functions and do all kinds of tasks including reading and writing

of data, writing of citations, data manipulations and plotting.

Typical R functions have some 'input' information or objects, which are then converted into some 'output' information or objects. The rules of syntax for creation of functions in R are specifically designed to facilitate collaboration and extension. Since R is 'open source,' the underlying code (however complicated) for all R functions is readily available for extension and /or modification. For example, typing 'matrix' prints the internal R code for the R function matrix to the screen, ready for modification. Of course, more useful idea is to type '?matrix' (no spaces, lower case m) to get the relevant user's manual page for using the function 'matrix'.

The true power of R can be exploited if one knows some basic rules of syntax for R functions, even if one may never write a new function. After all, hundreds of functions are already available for most numerical tasks in basic R and thousands more are available in various R packages. It is useful to know (i) how to create an object containing the output of a funtion function already in the memory of R, and (ii) the syntax for accessing the outputs of the function by the dollar symbol suffix. Hence, I illustrate these basic rules using a rather trivial R function for illustration purposes.

```
# R snippet 1.6.1 explains inputs /outputs of R functions.
myfunction=function(inp1,inp2, verbos=FALSE)
{ #function code begins with a curly brace
if(verbos) {print(inp1); print(inp2)}
out1=(inp1)^2 #first output squares the first input
out2=sin(inp2)#second output computes sin of second input
if(verbos) {print(out1); print(out2)}
list(out1=out1, out2=out2)#syntax for listing outputs
} #function code ENDS with a curly brace
```

The reader should copy and paste all the lines of snippet 1.6.1 into R and wait for the R prompt. Only if the function is logically consistent, R will return the prompt. If some errors are present, R will try to report to the screen those errors.

The first line of the function code in snippet 1.6.1 declares its name as 'myfunction' and lists the two objects called inp1 and inp2 as *required* inputs. The function also has an *optional* logical variable 'verbos=FALSE' as its third input to control the printing with a default setting (indicated by the equal symbol) as FALSE. All input objects must be listed inside parentheses (not brackets) separated by commas.

Since a function typically has several lines of code we must ask R to treat all of them together as a set. This is accomplished by placing them inside two curly braces. This is why the second line has left curly brace ({).

The third line of the snippet 1.6.1 has the command: 'if(verbos) print(inp1); print(inp2)'. This will optionally print the input objects to the screen if 'verbos' is 'TRUE' during a call to the function as 'myfunction(inp1,inp2, verbos=TRUE)'.

Now we enter the actual tasks of the function. Our (trivial) function squares the first input 'inp1' object and returns the squared values as the first output as an output object named 'out1'. The writer of the function has full freedom to name the objects (within reason so as not to conflict with other R names) The code "out1=inp1 2' creates the 'out1' for our function.

The function also needs to find the sine of second 'inp2' object and return the result as the second output named 'out2' here. This calculation is done here (still within the curly braces) by the code on the fifth line of the snippet 1.6.1. line 'out2=sin(inp2)'.

Once the outputs are created, the author of the R function must decide what names they will have as output objects. The author has the option to choose an output name different from the name internal to the function. The last but one line of a function usually has a 'list' of output objects. In our example it is 'list(out1=out1, out2=out2)'. Note that it has strange repetition of names. This is actually not strange, but allows the author complete freedom to name objects inside his function. The 'out1=out1' simply means that the author does not want to use an external name distinct from the internal name. For example, if the author had internally called the two outputs as o1 and o2, the list command would look like 'list(out1=o1, out2=o2)'. The last line of a function is usually a curly brace (}).

Any user of this function then 'calls' the R function 'myfunction' with the R command "myout=myfunction(in1,in2)". This command will not work unless the function object 'myfunction' and input objects inp1 and inp2 are already in the current memory of R. That is, the snippet 1.6.1 must be loaded and R must give hack the prompt before we can use the function.

Note that the symbol 'myout' on the left hand side of this R command is completely arbitrary and chosen by the user. Short and unique names are advisable. If no name is chosen, the command "myfunction(inp1,inp2)" will generally report certain outputs created by the function to the screen. By simply typing the chosen name 'myout' with carriage return, sends the output to the screen.

By choosing the name 'myout' we send the output of the function to an object bearing that name inside R. Self-explanatory names are advisable. More important, the snippet shows how to access the objects created by any existing R function for further manipulation by anyone by using the dollar symbol as a suffix followed by the name of the output object to the name of the object created by the function.

```
# R snippet 1.6.1b illustrates calling of R functions.
#Assume myfunction is already in memory of R
inp1=1:4 #define first input as 1,2,3,4
inp2=pi/2 #define second input as pi by 2
myout=myfunction(inp1,inp2)#creates object myout from myfunction
myout # reports the list output of myfunction
myout$out1^2  #^2 means raise to power 2 the output called out1
13*myout$out2#compute 13 times the second output out2
```

The snippet 1.6.1b illustrates how to create an object called 'myout' by using the R function called 'myfunction.' The dollar symbol attached to 'myout' then allows complete access to the outputs out1 and out2 created by 'myfunction.'

The snippet illustrates how to compute and print to the screen the square of the first output called 'out1'. It is accessed as an object under the name 'myout$out1,' for possible further manipulation, where the dollar symbol is a suffix. The last line of the snippet 1.6.1b shows how to compute and print to the screen 13 times second output object called 'out2' created by 'myfunction' of the snippet 1.6.1. The R output follows:

```
> myout # reports the list output of myfunction
$out1
[1]  1  4  9 16

$out2
[1] 1

> myout$out1^2  #^2 means raise to power 2 the output called out1
[1]   1  16  81 256
> 13*myout$out2#compute 13 times the second output out2
[1] 13
```

The rules of syntax in writing R functions are fairly simple and intuitive.

In any case, tens of thousands of (open source) functions have already been written to do almost any task spread over two thousand R packages. The reader can freely modify any of them (giving proper credit to original authors).

Accessing outputs of R functions with dollar symbol convention. The point to remember from the above illustration is that one need not write down the output of an R function on a piece of paper and type it in as an input to some other R function. R offers a simple and standardized way (by using the dollar symbol and output name as a suffix after the symbol) to cleanly access the outputs from any R function and then use them as inputs to another R function or manipulate collect, and report them as needed.

The exponential growth and popularity of R can be attributed to various convenient facilities to build on the work of others. The standardized access to function outputs by using the dollar symbol is one example of such facility in R. A second example is how R has standardized the non-trivial task of package writing. Every author of a new R package must follow certain detailed and strict rules describing all important details of data, inputs and outputs of all included functions with suitably standardized look and feel of all software manuals. Of course, important additional reasons facilitating world-wide collaboration include the fact that R is OOL, free and open source.

R packages consist of collections of such functions designed to achieve some specific tasks relevant in some scientific discipline. We have already encountered 'base' and 'contrib' packages in the process of installation of R. The 'base' package has the collection of hundreds of basic R functions belonging to the package. The "contrib" packages have other packages and add-ons. They are all available free and on demand as follows. Use the R-Gui menu called 'Packages' and from the menu choose 'install packages.' R asks you to choose a local comprehensive R archive network (CRAN) mirror or "CRAN mirror". Then R lets you "Install Packages" chosen by name from the long alphabetical list of some two thousand names. The great flexibility and power of R arises from the availability of these packages.

The users of packages are requested to give credit to the creators of packages by citing the authors in all publications. After all the developers of free R packages do not get any financial reward for their efforts. The 'citation' function of the 'base' package of R reports to the screen detailed citation information about citing any package.

Since there are over two thousand R packages available at the R website, it is not possible to summarize what they do. Some R packages of interest to me are described at my web page at:

`http://www.fordham.edu/economics/vinod/r-lang.doc`

My students find this file containing my personal notes about R (MS Word file some 88 pages long) as a time-saving devise. They use the search facility of MS Word on it.

One can use the standard *Google* search to get answers to R queries, but this can be inefficient since it catches the letter R from anywhere in the document. Instead, I find that it is much more efficient to search answers to any and all R-related questions at:

`http://www.rseek.org/`

1.7 R Object Types and Their Attributes

R is OOL with various types of objects. We note some types of objects in this section. (i) vector objects created by combining numbers separated by commas are already encountered above.

(ii) 'matrices' are objects in R which generalize vectors to have two dimensions. The 'array' objects in R can be 3-dimensional matrices.

(iii) 'factors' are categorical data objects.

Unlike some languages, R allows object to be mixed type with some vectors numerical and other vectors with characters. For example in stock market data, the Ticker symbol is a character vector and stock price is a numerical vector.

(iv) 'Complex number objects.' In mathematics numerical vectors can be complex numbers with $i = \sqrt{(-1)}$ attached to the imaginary part.

(v) R objects called 'lists' permit mixtures of all these types of objects.

(vi) Logical vectors contain only two values: TRUE and FALSE. R does understand the abbreviation T and F for them.

How does R distinguish between different types of objects? Each object has certain distinguishing 'attributes' such as 'mode' and 'length.' It is interesting that R can convert some objects by converting their attributes. For example, R can convert the numbers 1 and 0 into logical values TRUE and FALSE and vice versa. The 'as.logical' function of R applied to a bunch of zeros and ones makes them a corresponding bunch of TRUE and FALSE values. This is illustrated in Sec. 1.7.1

1.7.1 *Dataframe Matrix and Its Summary*

We think of matrix algebra in this book as a study of some relevant information. A part of the information can be studied in the form of a matrix. The 'summary' function is a very fundamental tool in R.

For example, most people have to contend with stock market or financial data at some point in their lives. Such data are best reported as 'data frame' objects in R, having matrix-like structure to deal with any kind of 'data matrices.' In medical data the sex of the patient is a categorical vector variable, patient name is a 'character' vector whereas patient pulse is a numerical variable. All such variables can be handled satisfactorily in R. The matrix object in R is reserved for numerical data. However, many operations analogous to those on numerical matrices (e.g., summarize the data) can be suitably redefined for non-numerical objects. We can conveniently implement all such operations in R by using the concept of data frame objects invented in the S language which preceded R.

We illustrate the construction of a data frame object in R by using the function 'data.frame' and then using the 'summary' function on the data frame object called 'mydata'. Note that R does a sensible job of summarizing the mixed data frame object containing character, numerical, categorical and logical variables. It also uses the 'modulo' function which finds the remainder of division of one number by another.

```
# R program snippet 1.7.1.1 is next.
x=c("In", "God", "We",  "Trust", ".")
x; print(x, quote=FALSE)
y=1:5 #first 5 numbers
z=y%%2  #modulo division of numbers by 2
# this yields 1,0,1,0,1 as five numbers
a=as.logical(z)
zf=factor(z)#create a categorical variable
data.frame(x,y,z,a, zf)
#rename data.frame as object mydata
mydata=data.frame(x,y,z,a, zf)
summary(mydata)
length(mydata) #how many rows?

> data.frame(x,y,z,a, zf)
      x y z      a zf
1     In 1 1   TRUE  1
```

```
2    God 2 0 FALSE  0
3     We 3 1  TRUE  1
4  Trust 4 0 FALSE  0
5      . 5 1  TRUE  1
> mydata=data.frame(x,y,z,a, zf)
> summary(mydata)
       x              y            z              a          zf
    .  :1    Min.    :1    Min.    :0.0    Mode :logical    0:2
  God  :1    1st Qu. :2    1st Qu. :0.0    FALSE:2          1:3
  In   :1    Median  :3    Median  :1.0    TRUE :3
  Trust:1    Mean    :3    Mean    :0.6    NA's :0
  We   :1    3rd Qu. :4    3rd Qu. :1.0
             Max.    :5    Max.    :1.0
> length(mydata)
[1] 5
```

The 'summary' function of R was encountered in the previous section
(Sec. 1.5). It is particularly useful when applied to a data frame object.
The 'summary' sensibly reports the number of times each word or symbol is
repeated for character data, the number of True and False values for logical
variables, and the number of observations in each category for categorical
or factor variables.

Note that the length of the data frame object is reported as the number
of rows in it. Each object comprising the data frame also has the exact
same length 5. Without this property we could not have constructed our
data frame object by the function 'data.frame.' If all objects are numerical
R has the 'cbind' function to bind the columns together into one matrix
object. If objects are mixed type with some 'character' type columns then
the 'cbind' function will convert them all into 'character' type before binding
the columns together into one character matrix object.

We now illustrate the use of 'as.matrix' and 'as.array' functions using the
data frame object called 'mydata' inside the earlier snippet. The following
snippet will not work unless the earlier snippet is in the memory of R.

```
# R program snippet 1.7.1.2 (as.matrix, as.list) is next.
#R.snippet Place previous snippet into R memory
#mydata=data.frame(x,y,z,a, zf)
as.matrix(mydata)
as.list(mydata)
```

Note from the following output that 'as.matrix' simply places quotes around the elements of the data frame object. The 'as.list' converts the character vector 'x' into a 'factor' arranges them alphabetically as 'levels'.

```
> as.matrix(mydata)
        x       y   z   a       zf
[1,] "In"     "1" "1" " TRUE" "1"
[2,] "God"    "2" "0" "FALSE" "0"
[3,] "We"     "3" "1" " TRUE" "1"
[4,] "Trust"  "4" "0" "FALSE" "0"
[5,] "."      "5" "1" " TRUE" "1"

> as.list(mydata)
$x
[1] In     God    We     Trust  .
Levels: . God In Trust We

$y
[1] 1 2 3 4 5

$z
[1] 1 0 1 0 1

$a
[1]  TRUE FALSE  TRUE FALSE  TRUE

$zf
[1] 1 0 1 0 1
Levels: 0 1
```

Unfortunately, having to know the various types of objects available in R may be discouraging or seem cumbersome to newcomers. Most readers can safely postpone a study of these things and learn these concepts only as needed. However, an understanding of classes of objects and related concepts described in this chapter will be seen to be helpful for readers wishing to exploit the true power of R.

Chapter 2

Elementary Geometry and Algebra Using R

2.1 Mathematical Functions

In Chap. 1 we discussed the notion of software program code (snippets) as R functions. The far more restrictive mathematical notion of a function, also called 'mapping,' is relevant in matrix algebra. We begin by a discussion of this concept.

Typical objects in mathematics are numbers. For example, the real number line $\Re^1 \in (-\infty, \infty)$ is an infinite set of all negative and positive numbers, however large or small. A function defined on $\Re^1$ is a rule which assigns a number in $\Re^1$ to each number in $\Re^1$. For example, $f : \Re^1 \to \Re^1$, where $f(x) = x - 12$ is well defined, and where the mapping notation $\to$ is used.

A function: Let X and Y be two sets of objects, where the objects need not be numbers, but more general mathematical objects including complex numbers, polynomials, functions, series, Euclidean spaces, etc. A function is defined as a rule which assigns one and only one object in Y to each object in X and we write $f : X \to Y$. The set X is called the 'domain' space where the function is defined. The set Y is called the 'range' or target space of the function.

The subset notation $f(X) \subset Y$ is used to say that 'X maps into Y.' The equality notation $f(X) = Y$ is used to say that 'X maps onto Y.' An image of an element of X need not be unique. For example, the image of the square root function can be positive or negative. The mapping is said to be one-to-one if each element of X has a distinct image in Y. If furthermore all elements of Y are images of some point in X, then the mapping is called one-to-one correspondence.

Composite Mapping or Function of a Function: We can also write a composite of two mappings:

$f : X \to Y$ and $g : Y \to Z$, as $g \circ f : X \to Z$.

This mapping is essentially a function g of function f stated in terms of the elemets of the sets as:

$$z = g[f(x)], x \in X, y \in Y, \text{ and } z \in Z. \tag{2.1}$$

2.2 Introductory Geometry and R Graphics

Chapter 1 discussed how to get started using R. The graphics system of R is particularly powerful in the sense that one can control small details of the plot appearance. This chapter hopes to make learning of geometry both a challenge and a 'hands on' experience, and therefore fun.

We begin with a simple plot of the Cartesian coordinate system using R. Recall from Sec. 1.6 that along each line of code, R ignores all text appearing after the # symbol. Section 1.5 item v described this. The snippet 2.2.1 uses the hash symbols to provide comments and remind ourselves and the reader what the nearby commands are doing. The reader should read the snippet comments and use the snippet as a template for using R graphics.

```
# R program snippet 2.2.1 plots coordinate system.
#first plot the origin and set up limits of axes
plot(0,0, xlim=c(-6,6), ylim=c(-6,6),
#now provide heading for the plot
main="Cartesian Coordinate System")
#plot command ended, Now Axes tick marks
Axis(c(-6,6),at=c(-6,6),side=1)
Axis(c(-6,6),at=c(-6,6),side=2)
grid() #explicity display the grid
abline(h=0) #draw horizontal axis line at the origin
abline(v=0)  #draw vertical line at the origin
# point P with x=2, y=4 as coordinates
text(2.3,4,"P") #place text P on the graph
text(-4.3,-6,"Q")# at specified coordinates. Note that for
#better visibility, horizontal coordinate is 0.3 units away.
arrows(0,0,2,4, angle=10)#command to draw vector as arrows.
arrows(0,0,-4,-6, angle=10)# with angle of the arrow tip
```

If one copies the above code into R, the Fig. 2.1 representing a two-dimensional plane is obtained.

In a Euclidean space the 'Real line' is represented by all real numbers in the infinite range represented by the open interval: $(-\infty, \infty)$. Parts of the real line falling within the limits of the graph are seen at the horizontal line (axis) passing through the values when $x = 0$ and $y = 0$ (i.e., the origin at $(0, 0)$ coordinate values). The figure represents an illustration of a two-dimensional Euclidian space split into four quadrants along four directions. Figure 2.1 also shows a grid representing the Cartesian coordinate system. We display two points P and Q in the North-East and South-West quadrants of the system.

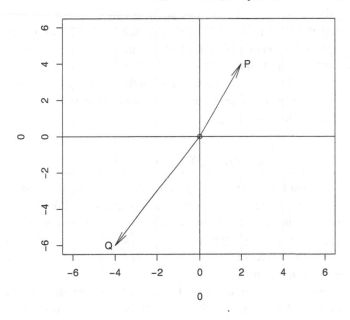

Cartesian Coordinate System

Fig. 2.1   The Cartesian coordinate system illustrated with point P(2, 4) at $x = 2, y = 4$ as coordinates in the North-East quadrant and point Q(−4, −6) in the South-West quadrant.

Figure 2.1 also shows arrows representing vectors $P = (2, 4)$ and $Q = (-4, -6)$ containing the two coordinates in a fixed order whereby x coordinate comes first. The arrows are drawn from the origin and represent the displacement starting at the origin moving x units on the horizontal axis and y units on the vertical axis.

More generally, the x axis can be the real line in the infinite open interval mentioned above. Similarly, the y axis can also be the real line $\Re$ placed at $90°$. Any point in this coordinate system formed by the two axes is represented by an 'ordered pair' of numbers (x, y), where it is customary to state the x coordinate value first. A two-dimensional space is represented by the Cartesian product denoted by the symbol $(\otimes)$ representing a set of ordered pairs of a set of x coordinate values with a set of y coordinate values:

$$\Re \otimes \Re = \{(x, y) : x \in \Re, y \in \Re\} = \Re^2. \tag{2.2}$$

Although the geometry might be difficult to represent on a piece of paper beyond two or three dimensions, nothing can stop us from imagining longer ordered values containing 3, 4 or any number n of coordinate axes. For example, $(1,1,1,1)$ as an ordered set of 4 numbers can represent the unit point in a 4 dimensional space.

Thus we can think of n dimensional Euclidean space. Now define the Euclidean distance between point P with coordinates $(p_1, p_2, \ldots p_n)$ and point Q with coordinates $(q_1, q_2, \ldots, q_n)$ as:

$$d(P, Q) = \sqrt{\Sigma_{i=1}^n (p_i - q_i)^2}. \tag{2.3}$$

Writing our own functions is a very useful facility in R to reduce tedious repetitive tasks in a compact, efficient and accurate way. We have encountered 'myfunction' in 1.6.1 for squaring one input and finding the sine of the second input. Here we illustrate a less trivial function. Section 5.7 has an R function with dozens of lines of code to compute answers to all types of payoff decision problems.

Let us write a general R function to compute the Euclidean distance. In the snippet 2.2.2 we name it 'Eudist'. It has two vectors 'p,q' as input stated by the code 'function(p,q)'. Recall that any R function always begins with left curly brace and ends with a right curly brace. The 'return' command here is used instead of the 'list' command to indicate the outputs of 'myfunction' in 1.6.1. The R experts discourage the use of 'return' method to end a function, which works only if exactly one output is produced. When only one output, the computed Euclidean distance, is produced, ending with 'return' has the advantage that no output name is produced. Hence the output is accessed without any dollar symbol. In the snippet 2.2.2 our function returns the Euclidean distance between the two input vectors as its output. The function is fully created in the first three lines of code of the snippet. The last three lines compute the distance for 1 to 3 dimensional vectors as inputs.

```
# R program snippet 2.2.2 is next.
Eudist = function(p,q) { #function begins here
ou1=(p-q)^2  #compute squared differences
return(sqrt(sum(ou1)))} #function ends here
Eudist(1,6) #one-dimensional distance
Eudist(c(2,4),c(-4,-6)) #2 dimensional distance
Eudist(c(2,3,4),c(2,8,16)) #3 dimensional distance
```

Recall the Figure 2.1 with arrows representing vectors $P = (2, 4)$ and $Q = (-4, -6)$. The snippet 2.2.2 finds the Euclidean distance between these two points to be 11.66190. The output of the above code snippet is given next.

```
> Eudist(1,6) #one-dimensional distance
[1] 5
> Eudist(c(2,4),c(-4,-6)) #2 dimensional distance
[1] 11.66190
> Eudist(c(2,3,4),c(2,8,16)) #3 dimensional distance
[1] 13
```

Given two points P and Q defined above, their convex combination for the n-dimensional case is the set of points $x \in \Re^n$ obtained for some $\lambda \in [0, 1]$ so that $x = \lambda P + (1 - \lambda)Q$ or more explicitly $x = \lambda p_1 + (1 - \lambda)q_1 + \ldots \lambda p_n + (1 - \lambda)q_n$.

2.2.1 *Graphs for Simple Mathematical Functions and Equations*

We have seen the plots of points similar to P and Q in Fig. 2.1 by using the 'plot' function of R. Now, we turn to plots for mathematical functions sometimes denoted as $y = f(x)$, where the equality sign represents an explicit equation defining y from any set of values of x. The function of two variables may be stated implicitly as $f(x, y) = 0$. Some trigonometric functions (such as $tan(\theta)$) are used in this subsection for handling angles.

Consider the equation $x + y = 12$, when $x = 1, 2, \ldots, 6$, the first six numbers. The corresponding $y = 11, 10, 9, 8, 7, 6$. are implied by the equation $x + y = 12$. The following R commands will produce the left panel of Fig. 2.2. The right panel of Fig. 2.2 represents the function $y = 2x$ produced by the code in the lower part of the following snippet.

```
# R program snippet 2.2.1.1 plots two lines.
par(mfrow=c(1,2))#plot format to have 1 row and 2 columns
x=1:6  #enter a sequence of numbers 1 to 6 as x
y=(12-1:6) #enter the y sequence from eq x+y=12
#plot command: main=main heading,
#typ="l" asks R to draw a line type plot
plot(x,y, main="Plot of Equation x+y=12", typ="l")
x2=2*x
plot(x,x2, main="Plot of Equation y=2x", typ="l",
ylab="y") #label y axis as y not x2
par(mfrow=c(1,1))# reset plot format to 1 row 1 column
```

The entire snippet(s) can be copied and pasted into R for implementation. Note from Fig. 2.2 that R automatically chooses the appropriate scaling for the vertical axis.

Angle between two straight lines:

If $y = a_1 x + b_1$ and $y = a_2 x + b_2$ are equations of two lines then $\tan\theta = (a_2 - a_1)/(1 + a_1 a_2)$ gives the angle through which the first line must be rotated to make it parallel to the second line.

For the example of two lines plotted in Fig. 2.2 we have $y = 12 - x, b_1 = 12, a_1 = -1$ and $y = 2x, b_2 = 0, a_2 = 2$, respectively. Now $\tan\theta = 3/(-1)$. $\tan^{-1}\theta = -1.249046$ radians or -71.56505 degrees computed by using the R function 'atan' for inverse tangent or arc tangent with the command 'atan(-3)*57.29578'. We can visualize the two curves together so that the angle between them can be visually seen. Hence, we use the following snippet.

```
# R program snippet 2.2.1.2 is next.
#R.snippet, assuming previous snippet is in memory
plot(x,y, main="Both x+y=12 and y=2x Lines",
ylim=c(0,12),xlim=c(0,7), typ="l")
#above sets up scaling to include both lines
abline(x,x2, ylab="y") #draws the second line
text(0.5,1, "y=2x")
```

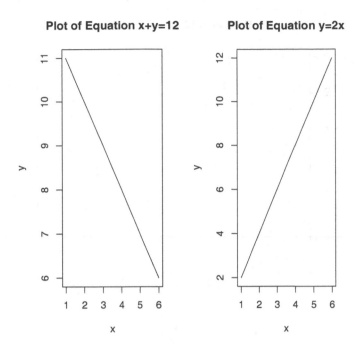

Fig. 2.2   Simple equation plots in R

text(2,10.5,"y=12-x")

The above snippet 2.2.1.2 shows that it takes a little work to get two lines on the same plot in R. We have to set up the plotting region carefully. Note that the second line is drawn by the function 'abline' rather than 'plot.' Another way to do the same (not seen in the snippet) is to use the function 'curve(2*x,add=TRUE),' where the argument 'add=TRUE' tells R to add a function to the existing plot using existing values on the horizontal axis. The snippet also shows how to insert text in the plot by specifying the coordinate points. A little trial and error is needed to get it to look right.

2.3   Solving Linear Equation by Finding Roots

Matrix algebra provides powerful tools for solving one or more equations. By way of introduction, it helps the intuition to try to understand the geometry behind the notion of solving a linear equation $y = f(x) = 0$.

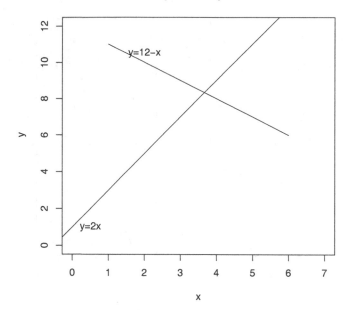

Fig. 2.3   Two mathematical equations plotted together to visualize the angle between them.

Solving it means that we want to find the correct number of values of x for which the equation holds true. Geometrically, $y = 0$ is a line and the number of solutions equals the number of times $f(x)$ touches or crosses the line.

Note that the angle between the two lines is only 45 degrees according to the formal computation. The $\tan(\theta) = 1$ is calculated in the snippet. The angle looks visually larger, since the range of the vertical axis [-4,4] is twice as wide as that of the horizontal axis [-1,3].

R program snippet **2.3.1** is next.

```
x=seq(-1,3,by=0.1)
y1=2*x-3
y2=-3*x+2; y3=rep(0,length(x))
plot(x,y1, main="Solving y=2x-3 and y=-3x+2 Lines",
ylim=c(-5,5),xlim=c(-1,3), typ="l", ylab="f(x)")
```

```
text(0,-4, "y= 2 x - 3")
#above sets up scaling to include both lines
lines(x,y2) #draws the second line
text(0, 4, "y= - 3 x + 2")
lines(x, y3) #draws axis
grid(nx=2,ny=0, lwd=2) #wide vertical line at x=1
a1=2; a2=-3
tanTheta=(a2-a1)/(1+a1*a2)
theta=atan(tanTheta)*57.29578; theta
```

Now note that the solution of the equation $f(x) = 2x - 3 = 0$ is at $x = 3/2$, a positive value larger than 1 compared to the solution of the equation $f(x) = -3x + 2 = 0$ at $x = 2/3$, a value smaller than 1. Geometrically they are seen in Fig. 2.4 at two points where the line $y = 0$ meets the respective lines.

The equations need not be simply mathematical exercises, but can mean profound summaries of the macro economy. For example, y can represent the 'real' or inflation-adjusted interest rates and x can represent the gross domestic product (GDP). Such an intersecting pair of equations is known as the 'linear IS-LM model' in macroeconomics. The line top left to bottom right is known a the 'IS' curve representing investments and savings behavior in response to interest rates. The line from bottom left to top right is called the 'LM' curve equating aggregate money demand with money supply. The point of intersection is known as the equilibrium solution of interest rate and GDP. Note that the intersection (equilibrium) occurs at a negative value of the interest rates (vertical axis). Of course, no bank can charge negative interest rate as a practical matter. Hence monetary policy cannot get the economy represented by Fig. 2.4 out of disequilibrium (recession). Only fiscal policy with deficit spending which will inject inflation can achieve an equilibrium. The point is that innocent looking equations can mean a great deal.

2.4 Polyroot Function in R

Most students are familiar with finding the roots of a quadratic equation, but do not know how to find the roots of equations involving higher order polynomials. This section shows that R removes the drudgery behind the root-finding task.

Both y=2x–3 and y=–3x+2 Lines

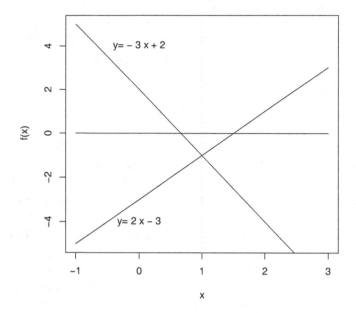

Fig. 2.4   Solving two linear equations $f(x) = 2x - 3 = 0$ and $f(x) = -3x + 2 = 0$.

The general formula for the quadratic equation is

$$f(x) = a\,x^2 + b\,x + c = 0, \tag{2.4}$$

where a, b, c are known constants. The roots of Eq. (2.4) are real numbers only if $\rho = b^2 - 4\,a\,c \geq 0$, otherwise the roots are imaginary numbers. Consider a simple example where $a = 1, b = -5, c = 6$ having real roots. The general formula for two roots is:

$$x = \frac{-b \pm \sqrt{(b^2 - 4\,a\,c)}}{2a}. \tag{2.5}$$

Substituting $a = 1, b = -5, c = 6$ in Eq. (2.5) we find that $\rho = 25 - 24 = 1$, and the two roots are $x_1 = [(5 + 1)/2] = 3, x_2 = [(5 - 1)/2] = 2$. That is the quadratic Eq. (2.4) can be factored as $(x - 3)(x - 2)$.

The following important lessons from this simple high-school algebra exercise generalize to higher order polynomials and are relevant for matrix theory of this book.

1] The number of roots (=2, here) equals the highest power of x in Eq. (2.4).

2] The finding of roots is a non-trivial task in old-fashioned calculator methods, especially since roots can often involve imaginary numbers of the type $x + iy$ where $i = \sqrt{(-1)}$.

The roots of a polynomial are available in R by calling the function 'polyroot'. R defines a polynomial of degree $n - 1$ as $f(x) = z1 + z2 * x + ... + zn * x^{(n-1)}$, where the constant (intercept) term $z1$ comes first, the coefficient $z2$ of the linear term (coefficient of x) comes next, the coefficient $z3$ of the quadratic term (x^2) comes next, and so on.

For example, if you want R to solve $f(x) = 2x - 3 = 0$, the R command is 'polyroot(c(-3,2))'. Now R responds as: '[1] 1.5+0i,' which is a complex number with imaginary part zero. The '+0i' part of the R response can, of course, be ignored. Then the root is understood to be: $3/2 = 1.5$. Clearly, it is not necessary to use 'polyroot' to solve simple linear equations. I am using the simple example so the reader can see that R gets the right answer and let the reader see the somewhat sophisticated (peculiar to those unfamiliar with complex numbers) way in which the input to and the outputs (answers) from R are stated in R. It is unfortunate that R does not automatically delete the imaginary part when it is zero (when the answers are real numbers).

Of course, the real power of the 'polyroot' function of R arises from its ability to numerically solve for the roots of high order polynomials. For example, consider the quintic polynomial $(x + 1)^5 - 0$ with the respective coefficients obtained by the Binomial expansion as: (1, 5, 10, 10, 5, 1). It obviously has 5 roots, all of which are -1. The analogous quintic $(x - 1)^5 = 0$ has the same coefficients, except that they alternate in sign as (1,-5,10,-10,5,-1). The snippet 2.4.1 illustrates the solutions for both cases. The input to the polyroot function is simply the list of coefficients defined by the 'c' function Sec. 1.5 for various R tips, specifically in vi. The 'polyroot' function of R will report an 'Error' if 'c' is not used.

```
# R program snippet 2.4.1 is next.
polyroot(c(1,5,10,10,5,1))
polyroot(c(1,-5,10,-10,5,-1))
```

We have chosen this particular example to illustrate 'polyroot' because we already know the correct answer. Our aim here is to illustrate the use of the function and check that it is correctly finding the roots in a known case. Then R responds again with some '+0i' and some '-0i,' both to be ignored, as:

```
> polyroot(c(1,5,10,10,5,1))
[1] -1-0i -1+0i -1-0i -1-0i -1-0i
> polyroot(c(1,-5,10,-10,5,-1))
[1] 1+0i 1-0i 1-0i 1+0i 1+0i
```

Recall that in Chap. 1 Sec. 1.7 we defined attributes of vectors. It is possible to get rid of the '+0i' and '-0i' by using the 'as.numeric' function of R to coerce R into reporting a more readable set of (real number) roots as illustrated in the following snippet.

```
# R program snippet 2.4.2 is next.
p1=polyroot(c(1,5,10,10,5,1)) #define object p1
p2=polyroot(c(1,-5,10,-10,5,-1)) #define object p2
as.numeric(p1)
as.numeric(p2)
```

Finally, R responds without the '+0i' and '-0i,' as:

```
> as.numeric(p1)
[1] -1 -1 -1 -1 -1
Warning message:
imaginary parts discarded in coercion
> as.numeric(p2)
[1] 1 1 1 1 1
Warning message:
imaginary parts discarded in coercion
```

Note that 'polyroot' correctly finds the five roots of $(x+1)^5 = 0$ to be all -1 and the roots of $(x-1)^5 = 0$ to be all $+1$. The example is artificially designed to have only real numbers as roots. Since the polynomial roots are, in general, complex numbers, the roots found by the 'polyroot' function of R have the correct attribute of 'complex numbers' in the sense described in Sec. 1.7.

2.5   Bivariate Second Degree Equations and Their Plots

Let us begin with bivariate (with two variables x and y) first degree curves before turning to the second degree. A simple bivariate equation of the form $Ax + By + C = 0$, where at least one of the coefficients A, B is nonzero, is called a bivariate algebraic equation of the first degree. Geometrically, first degree equations always represents a straight line.

More generally, a bivariate algebraic equation of the second degree is defined as:

$$Ax^2 + Bxy + Cy^2 + Dx + Ey + F = 0, \qquad (2.6)$$

where at least one of the coefficients A, B, C is non-zero. This section shows that Eq. (2.6) can represents many kinds of figures. Later we shall use a matrix algebra concept called the 'determinant' allows us to know precisely when it represents two lines of only one line. This is one of the motivations for including this material in a book on matrix algebra.

The circle of radius R centered at the origin is a special case of the above equation defined by: $x^2 + y^2 = R^2$. The inequality $B^2 + C^2 - 4AD > 0$ must be satisfied for Eq. (2.6) to be a circle.

It is fun to be able to plot algebraic functions using R. Plotting the unit circle defined by the equation $x^2 + y^2 = 1$ in R requires us to first define the values of x on the horizontal axis by using a 'seq' function for creating a sequence of numbers between -1 to 1. For plotting purposes, R needs to know the value on the vertical axis y for each value on the horizontal axis x. Next, we need to choose the limits on the vertical axis from -1 to 1. These limits are given as argument to the R function 'plot' as 'ylim=c(-1,1)' in the following snippet 2.5.1.

The circle has to be defined from $\pm\sqrt{(1 - x^2)}$ defined by two segments. The first segment occurs when the root is positive and the second segment occurs when it is negative. We must define them separately for the purpose of plotting. Drawing a second line on the existing plot in R cannot be done by the 'plot' command, since 'plot' always draws a new plot. One needs to use the 'lines' function in R as in the following snippet. The ellipse as a compressed circle for the equation $x^2 + 4y^2 = 1$ is plotted in a similar fashion. We place them in left and right panels of the same figure by using the command 'par(mfrow=c(1,2))' as we did before in the snippet 2.2.1.1. The snippet 2.5.1 illustrates the use of the R function 'seq(from, to, by=)' to generate sequence of numbers for specified start and end and the value of the increment.

```
# R program snippet 2.5.1 Plots Circle and Ellipse.
par(mfrow=c(1,2)) #plot 1 row and 2 column plots
par(pty="s") #square plot area avoids visual distortions
x=seq(-1,1,by=0.01) #R understand that from=-1, to=1
```

```
plot(x,-sqrt(1-x^2),typ="l", ylim=c(-1,1),
main="Plot of Unit Circle", ylab="y")
lines(x,sqrt(1-x^2),typ="l") #draws second half
#now plot ellipse
plot(x,-0.5*sqrt(1-x^2),typ="l", ylim=c(-1,1),
main="Plot of An Ellipse", ylab="y")
lines(x,0.5*sqrt(1-x^2),typ="l")#second half
par(mfrow=c(1,1)) #reset to usual plots
par(pty="m") #reset to maximization of area
```

The standard form of an ellipse (compressed circle) is: $\frac{x^2}{a^2} + \frac{y^2}{b^2} = 1$. For example, $x^2 + 4y^2 = 1$ is an ellipse in our Fig. 2.5 here. Note that the circle is a special case of ellipse when $a = b$.

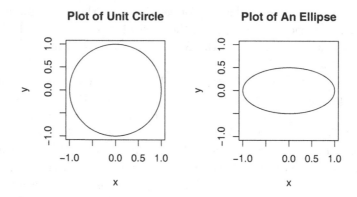

Fig. 2.5 Plots of second order equations for the unit circle $x^2 + y^2 = 1$ and an ellipse $x^2 + 4y^2 = 1$ in R.

The equation of hyperbola is: $\frac{x^2}{a^2} - \frac{y^2}{b^2} = 1$.

The equation of conjugate hyperbola is: $\frac{x^2}{a^2} - \frac{y^2}{b^2} = -1$. It involves imaginary numbers. Our figure Fig. 2.6 illustrates the hyperbola for the equation $x^2 - y^2 = 1$.

R program snippet **2.5.2** plots Hyperbola.
```
x=seq(-1,1,by=0.01)
plot(x,-sqrt(1+x^2),typ="l", ylim=c(-2,2),
main="Plot of A Hyperbola", ylab="y")
lines(x,sqrt(1+x^2),typ="l") #draws upper half
```

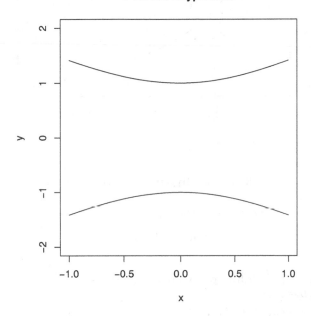

Fig. 2.6   Plots of a Hyperbola for the equation $x^2 - y^2 = 1$ in R.

A rectangular hyperbola is defined by the equation $xy = a$, where a is a constant. Clearly, it satisfies $y = a/x$. The following snippet produces Fig. 2.7 upon choosing $a = 5$ and a range for $x \in [1, 5]$.

R program snippet **2.5.3** plots rectangular hyperbola.
```
x=seq(0,5,by=0.01)
plot(x,5/x, typ="l", main="Rectangular Hyperbola")
```

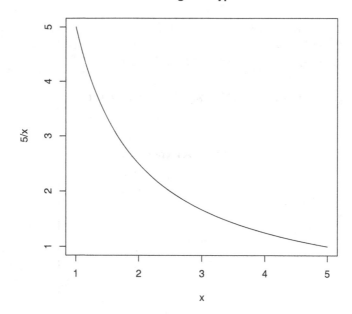

Fig. 2.7   Plot of a Rectangular Hyperbola for the equation $xy = 5$ in R.

The parabola is the graph of the equation

$$y = ax^2 + bx + c. \tag{2.7}$$

R program snippet **2.5.3** for two parabola plots.
```
x=seq(-1,1,by=0.01)
plot(x,3*x^2-2*x, typ="l",
main="Two Parabola Plots", ylab="y")
lines(x,2*x^2+2*x)
text(0.5,3,"a=2, b=2, c=0")
text(-0.7,1.1,"a=3, b= - 2, c=0")
polyroot((c(0,-2,3)) #[1] 0.0000000+0i 0.6666667+0i
#solutions are x=0 and x=2/3
polyroot((c(0,2,2)) #[1]   0+0i -1+0i
#solutions are x=0 and x=-1
```

Since both parabolas depicted in Fig. 2.8 are obtained by a second

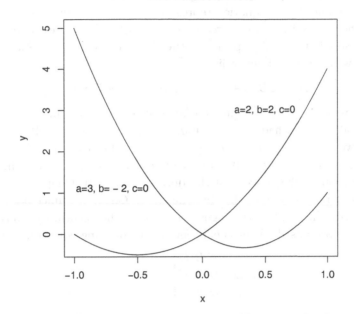

Fig. 2.8 Two plots of parabolas from the equation $y = ax^2 + bx + c$ displaying the effect of the size of $a = 2, 3$ and the sign of $b = -2, 2$.

degree polynomial (quadratic), they have two roots or two places where the line $y = 0$ intersects each parabola. Note that both parabola equations $f(x) = 3 * x^2 - 2 * x$ and $g(x) = 2 * x^2 + 2 * x$ have the intercept $c = 0$ in the notation of Eq. (2.7). Hence, when $x = 0$ we also have $y = 0$ for both parabolic curves. Therefore, $x = 0$ is one of the two solutions of each quadratic (by design). The point of intersection of the two curves in our Fig. 2.8 is at the point whose coordinates are: (0, 0). For this example, the point of intersection having $x = 0$ also represents the first solution (or root) of both equations. The second root can be obtained analytically as: $x = 2/3$ for the curve representing the equation $f(x) = 0$, and at $x = -1$ for the equation $g(x) = 0$. Verify that one can also obtain these pairs of roots by using the R function 'polyroot,' as indicated in snippet 2.5.3.

Although parabola, circle, ellipse and hyperbola are all defined by second degree equations, not all second degree equations imply one of these shapes. Indeed some second degree equations might represent two straight lines. For example, equation $x^2 - 2xy + y^2 - 25 = 0$ represents two lines: $x - y + 5 = 0$ and $x - y - 5 = 0$. It is generally tedious to figure out whether the given second degree equation represents one curve or two lines.

Let the original second degree equation Eq. (2.6) be rewritten after inserting an additional multiplier 2 to the three coefficients, (coefficients of B, D, E) without loss of generality, as:

$$Ax^2 + 2Bxy + Cy^2 + 2Dx + 2Ey + F = 0. \tag{2.8}$$

The 'no loss of generality' claim can be seen by the following example. If the original equation had the term $4xy$, we simply choose $B = 2$.

Now we can use matrix algebra to write quickly the condition under which equation Eq. (2.8) represents two lines. We postpone the definition and discussion of the concept of the determinant until Chap. 6 after we have defined related matrix algebra concepts in Chap. 4. Suffice it to state that powerful theory of matrix algebra has established centuries ago that if the determinant based on the coefficients of the defining equation satisfies:

$$det \begin{bmatrix} A & B & D \\ B & C & E \\ D & E & F \end{bmatrix} = 0, \tag{2.9}$$

then Eq. (2.8) represents two straight lines instead of only one curve.

For example, if we choose A to F as (1,2,5,3,6,9), or (2,-2,2,4,-4,-17), respectively, they give a zero determinant. This implies that both examples must be geometrically represented by two straight lines instead of single curves.

The numerical evaluation of a determinant in R begins with the definition of a matrix, which in turn must begin with a column-wise (not row-wise) list of all its elements. For the example of the second degree equation above the column-wise list of all elements is $(A, B, D, B, C, E, D, E, F)$. For example, the R commands to evaluate the determinant for the case when A to F are (1,2,5,3,6,9) are in the next snippet 2.5.4.

```
# R program snippet 2.5.4 has matrix and determinant.
x=c(1,2,3,2,5,6,3,6,9)
mym=matrix(x,nrow=3,ncol=3)
mym #print the matrix mym
det(mym) #compute and print the value of determinant
```

The notation 'mym' is used to define the 3×3 matrix from the list of numbers. The arguments 'ncol' specifies the number of columns and similarly the arguments 'nrow' specifies the number of rows in a rather intuitive notation.

We remind the reader that we have already mentioned in Sec. 1.6 and Sec. 1.5 that according to the syntax of R any lists of numbers must be separated by commas and must be preceded by a 'c' in R, where the 'c' is an important R 'function' presumably coming from the first character of the word 'combine'. If one types the command 'x=(1,2,3,2,5,6,3,6,9)', (forgetting to include the c) in the above snippet 2.5.4, R will refuse to accept it as a definition of x and produce the error report:

```
Error: unexpected ',' in "x=(1,"}.
```

The obviously named function 'matrix' of R creates a matrix from the given list of numbers with specified number of rows and columns. The snippet 2.5.4 also shows that the R function 'det' (similarly obviously named) computes the determinant. The output of the above set of R commands is:

```
> mym #print the matrix mym
     [,1] [,2] [,3]
[1,]   1    2    3
[2,]   2    5    6
[3,]   3    6    9
> det(mym) #compute and print the value of determinant
[1] -4.996004e-16
```

Note that R reports matrices by indicating the row and column numbers as headings. The intuitive '[i,]' means the i-th row and '[,j]' is for the j-th column. The scalar output precedes with [1]. The last line of the R output above means that the determinant is $-4.996004 * 10^{-16}$, in the engineering notation, or almost zero.

Any software has to have some formal notation to understand its components. The point to remember is that the notation in R is mostly very intuitive.

If one tries to create a matrix out of inadequate list of numbers, R has the (unfortunate?) habit of recycling the numbers till it is complete. For example if we try to create a 6×2 matrix out of first five numbers by using the command 'matrix(1:6,6,2)', here is how R recycles them and includes a warning that it has done recycling.

```
> matrix(1:5,6,2)
     [,1] [,2]
```

```
[1,]    1    2
[2,]    2    3
[3,]    3    4
[4,]    4    5
[5,]    5    1
[6,]    1    2
Warning message:
In matrix(1:5, 6, 2) :
   data length [5] is not a sub-multiple or multiple of the
   number of rows [6]
```

If one forgets the name of any R function, simply type its intuitive name preceded by '??.' For example typing '??matrix' into R gives a very long list of places where a word or phrase similar to 'matrix' appears inside all packages loaded in your version of R. The R package called 'multipol' by R. K. S. Hankin provides tools for manipulating multivariate polynomials.

Chapter 3

Vector Spaces

3.1 Vectors

We have already come across vectors in the Preface and in Chap. 1 and Chap. 2. For example, in Sec. 1.6 we define 'in1=c(2,4,6,99,5)' as a vector of five real numbers when we explain the 'c' (combine) function of R. We also noted in Sec. 1.7 that a vector in R can consist of logical values ('TRUE' and 'FALSE') or characters (in quotation marks) or imaginary (complex) numbers.

The 'field' of real numbers is usually denoted by $\Re$, while the field of complex numbers is denoted by $\mathscr{C}$. The entries of matrices are often shown as subsets of these fields.

Numerical vectors have a geometric representation. Figure 2.1 has two arrows starting from the origin to the points P and Q represented by the x and y coordinates of the two points, which in turn are represented by two vectors $P = (2, 4)$ and $Q = (-4, -6)$. Recall that these vectors are 'ordered' coordinate pairs, in the sense that the x value must be listed as the first among the two numbers inside the vectors.

Dimension of Space and Vector: In Sec. 2.2 we define a vector by an arrow from the origin. Any point $P = (x, y)$ in a two-dimensional Euclidean space has two coordinates which can be represented by a vector with two elements. Using Pythagoras' theorem the length of the point P from the origin is simply $\sqrt{(x^2 + y^2)}$. In general, a vector with n elements (coordinates) is represented in an n-dimensional space. Recall from Chap. 2 that Eq. (2.3) provides a formula to compute the Euclidean distance between two vectors containing ordered coordinates $(p_1, p_2, \ldots p_n)$ and $(q_1, q_2, \ldots, q_n)$.

41

3.1.1   *Inner or Dot Product and Euclidean Length or Norm*

The Euclidean inner product (dot product) of a vector with itself is simply the scalar (simple number) computed as the sum of squares (SuSq) of all elements in that vector. We postpone our discussion of (general) inner products possibly involving complex numbers and middle weighting matrices. The SuSq for vector $P = (p_1, p_2, \ldots p_n)$ is defined as:

$$\text{SuSq(P)} = \Sigma_{i=1}^{n}(p_i)^2. \tag{3.1}$$

The inner product (dot product) of a vector with another vector is also a scalar computed as the sum of products (SuPr) of all elements in a vector with the corresponding elements of another vector. After we define the concept of matrices, we can discuss the notion of inner products where the underlying matrix of 'quadratic /bilinear form' is the identity matrix. The Euclidean inner product of $P = (p_1, p_2, \ldots p_n)$ with another vector $Q = (q_1, q_2, \ldots, q_n)$ (sometimes denoted as $< p \bullet q >$) is defined as:

$$\text{SuPr(P, Q)} = < p \bullet q > == \Sigma_{i=1}^{n}(p_i q_i). \tag{3.2}$$

The Euclidean length of the arrow from the origin to the point representing a vector P is denoted by $\parallel P \parallel$ and is calculated by using the formula:

$$\parallel P \parallel = \sqrt{\Sigma_{i=1}^{n}(p_i)^2}. \tag{3.3}$$

Note that $\parallel P \parallel = \sqrt{\text{SuSq(P)}}$ holds true. That is the Euclidean inner product of a vector (with itself) is its squared Euclidean length.

Vector Norm: The length of a vector is also called the Euclidean norm. The 'norm' refers to size and in Euclidean geometry the size is often related to the length.

Now we illustrate these with examples using R for a 4-dimensional Euclidean space which exists only in our imagination. Let us randomly choose 4 numbers from out of the integers in the range 10 to 20 as our P coordinates and similarly choose our Q. In order to get identical result every time we run R, we set the seed of the random number generator by the function 'set.seed.' We use the 'sample' function with the argument 'size=4' in R to do the random selection. The 'length' function in R is *not* the Euclidean length, but just the length of the list or the number of elements in the vector. Thus for numerical vectors the dimension of the

\# R program snippet **3.1.1.1** for Euclidean lengths is next.

```
set.seed(978)
p=sample(10:20, size=4);p; length(p)
q=sample(10:20, size=4);q; length(q)
SuSqP=sum(p^2);SuSqP
SuSqQ=sum(q^2);SuSqQ
Euc.lenP=sqrt(SuSqP)
Euc.lenQ=sqrt(SuSqQ)
SuPr.PQ=sum(p * q)
rbind(SuSqP, SuSqQ, Euc.lenP, Euc.lenQ, SuPr.PQ)
```

The snippet 3.1.1.1 uses the 'rbind' function of R toward the end to bind the rows of computed answers after choosing self-explanatory names for R objects created during our computations. They are printed one below the other as a result of the 'rbind'.

```
> p=sample(10:20, size=4);p; length(p)
[1] 20 11 10 19
[1] 4
> q=sample(10:20, size=4);q; length(q)
[1] 10 20 15 12
[1] 4
> SuSqP=sum(p^2);SuSqP
[1] 982
> SuSqQ=sum(q^2);SuSqQ
[1] 869
> Euc.lenP=sqrt(SuSqP)
> Euc.lenQ=sqrt(SuSqQ)
> SuPr.PQ=sum(p * q)
> rbind(SuSqP, SuSqQ, Euc.lenP, Euc.lenQ, SuPr.PQ)
                [,1]
SuSqP      982.00000
SuSqQ      869.00000
Euc.lenP    31.33688
Euc.lenQ    29.47881
SuPr.PQ    798.00000
```

3.1.2 Angle Between Two Vectors, Orthogonal Vectors

Section 2.2.1 discusses angle between two straight lines. Here we are interested in the angle between two vectors. For graphical representation it is convenient to consider only two dimensional vectors defined as $P = (2, 4), Q = (5, 2)$. As before we can readily compute the Euclidean inner product (dot product) between these two vectors $< p \bullet q >$ and their Euclidean lengths $\| p \|$ and $\| q \|$, respectively.

R program snippet **3.1.2.1** for Angle Between Two Vectors.

```
#first plot the origin and set up limits
p=c(2,4); q=c(5,2)
SuSqP=sum(p^2)
SuSqQ=sum(q^2)
Euc.lenP=sqrt(SuSqP)
Euc.lenQ=sqrt(SuSqQ)
SuPr.PQ=sum(p * q)
cos.theta=SuPr.PQ/(Euc.lenP*Euc.lenQ)
theta=abs((acos(cos.theta) -pi/2 )*57.29578)
#angle in degrees instead of radians
segment.length=Euc.lenP*cos.theta
rbind(SuSqP, SuSqQ, Euc.lenP, Euc.lenQ,
SuPr.PQ, cos.theta, theta, segment.length)
plot(0,0, xlim=c(-1,6), ylim=c(-1,6),
main="Angle Between Two Vectors",xlab="x", ylab="y")
#plot command ended, Now Axes tick marks
#Axis(c(-6,6),at=c(-6,6),side=1)
#Axis(c(-6,6),at=c(-6,6),side=2)
grid()
abline(h=0) #draw horizontal axis line
abline(v=0)  #draw vertical line
# point P with x=2, y=4 as coordinates
text(2.1,4.1,"P") #place text on the graph
text(5.1,2.1,"Q")#at specified coordinates
arrows(0,0,2,4, angle=8) #arrow angle
arrows(0,0,5,2, angle=8)
text(3.4, 1.2, "A")
text(-.15, -.15, "O")
```

As the output of 3.1.2.1 from R indicates the inner product of these two vectors $< p \bullet q >= 18$, Euclidean lengths are Euc.lenP =4.4721360 and Euc.lenQ=5.3851648, respectively. We know from theory that the angle between the vector OP and OP in Fig. 3.1 satisfies the following relation:

$$\cos(\theta + 0.5\pi) = \frac{<\mathbf{p} \bullet \mathbf{q}>}{\| \mathbf{p} \| \| \mathbf{q} \|} \quad (3.4)$$

If we drop a perpendicular from point P on the line OQ it meets OP at point marked A in Fig. 3.1. The distance OA is known to equal $\| p \| \cos\theta$. The snippet 3.1.2.1 computes it as the object 'segment.length' and reports it to be 3.7139068.

The angle is obtained as arc-cosine or inverse cosine from the following definition:

$$\theta = \cos^{-1} \left[\frac{<\mathbf{p} \bullet \mathbf{q}>}{\| \mathbf{p} \| \| \mathbf{q} \|} \right] - \frac{\pi}{2}. \quad (3.5)$$

The angle in radians is $\pi/2$ or 90 degrees when the cosine is zero. In that case the vectors are said to be orthogonal or perpendicular. The actual angle in degrees between vectors $p = (2,4)$ and $q = (5,2)$ represented by arrows OP and OQ in Fig. 3.1 is 48.3664611, which is obtained upon multiplication of the angle in radians by 57.29578 in the R snippet 3.1.2.1.

```
                     [,1]
SuSqP          20.0000000
SuSqQ          29.0000000
Euc.lenP        4.4721360
Euc.lenQ        5.3851648
SuPr.PQ        18.0000000
cos.theta       0.7474093
theta          48.3664611
segment.length  3.3425161
```

Some mathematicians write the angle in Eq. (3.4) purely in terms of Euclidean lengths as:

$$\cos(\theta + 0.5\pi) = \frac{\| p \|^2 + \| q \|^2 - \| p - q \|^2}{2 \| p \| \| q \|}, \quad (3.6)$$

where The length of the line joining the points P and Q is

$|p - q| = \sqrt{(\Sigma_{i=1}^{n} [p_i - q_i]^2)} = \sqrt{(\Sigma_{i=1}^{n} [p_i^2 + q_i^2 - 2p_i q_i])}.$

It is easy to use R to note that both formulas yield exactly the same right hand side (RHS) denoted in the snippet above as 'cos.theta' (=0.7474093)

```
# R program snippet 3.1.2.1 for Euclidean Length cosine theta next.
p=c(2,4); q=c(5,2)
SuSqP=sum(p^2)
```

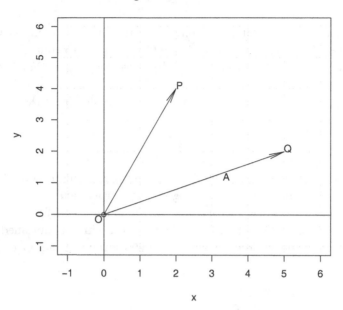

Fig. 3.1   Angle between two vectors OP and OQ

```
SuSqQ=sum(q^2)
pmq=p-q
SuSqpmq=sum(pmq^2)
Euc.lenP=sqrt(SuSqP)
Euc.lenQ=sqrt(SuSqQ)
RHS=(SuSqP+SuSqQ-SuSqpmq)/(2*Euc.lenP*Euc.lenQ);
RHS

> RHS
[1] 0.7474093
```

3.2   Vector Spaces and Linear Operations

We have come across the n dimensional Euclidean space $\Re^n$. The vector space is a subspace of $\Re^n$. The matrix theory does not need all properties of $\Re^n$. A vector space seeks to retain only those properties needed for linear operations on vectors. Linear operations are weighted sums. The totality of weighted sums with arbitrary choices of weights is called a vector space.

The key property according to Marcus and Minc (1964) is called '**closure:**' If a vector v is in a vector space V, then any linear combination of vectors in V is also in V.

Definition of Vector Space

Formally we can define a vector space V as a collection of $n \times 1$ vectors v satisfying the following properties: (1) If $v_1 \in V$ and $v_2 \in V$, then $v_1 + v_2 \in V$. (2) If a is any real number and $v \in V$ the $av \in V$. The choice $a = 0$ leads to the null vector with all elements zero and it must belong to every vector space.

3.2.1 *Linear Independence, Spanning and Basis*

Consider two vectors $P = (x, y)$ and $Q = (2x, 2y)$ and represent them by arrows from the origin. Since every coordinate of P is half the value of the coordinate of Q, the vector OP is simply half a segment of OQ. These vectors lie on top of each other and illustrate linearly dependent vectors.

Given P and Q, consider a set spanned (formulated) by the linear combination (weighted sum):

$$L(P, Q) = \{w_1 P + w_2 Q : w_1 \in \Re, \text{and} \, w_2 \in \Re\}. \tag{3.7}$$

When $P - (x, y)$ and $Q = (2x, 2y)$ verify that the spanning set $L(P, Q)$ is degenerate lying on one straight line in the direction OP or are 'collinear.'

Now let us define $P = (2, 4), Q = (5, 2)$, the two vectors depicted in Fig. 3.1. It is visually clear that these vectors are not collinear (linearly independent). In fact it can be shown that the spanning set from Eq. (3.7) based on these two vectors can indeed span the whole two dimensional Euclidean space $\Re^2$. Of course, the suitable choice of weights can involve large or small, positive or negative values. Next, we illustrate this idea.

Since we have been extending two-dimensional concepts to $\Re^n$, it helps to change the notation P, Q to p_1, p_2, and insert an additional subscript (1,2) for the ordered values within each vector. Thus, we rewrite $p_1 = (p_{11}, p_{12}) = (2, 4)$ and $p_2 = (p_{21}, p_{22}) = (5, 2)$. Next, we want to verify that an arbitrary point in the 2-dimensional space with coordinates $(99, -3)$ can be constructed from these two vectors with suitable weights w_1, w_2. We can think of the arbitrary point as right hand side values $(r_1, r_2) = (99, -3)$. As long as such weights exist for an arbitrary vector of (right hand) (r_1, r_2)

values, we are home.

Numerically finding the two weights w_1, w_2 is a different matter and will be discussed later. Here we are concerned with the existence. Finding weights amounts to solving the following system of two equations with these weights as unknowns.

$$p_{11}\, w_1 + p_{12}\, w_2 = r_1, \tag{3.8}$$

$$p_{21}\, w_1 + p_{22}\, w_2 = r_2. \tag{3.9}$$

Our first example had $P = (x, y)$ and $Q = (2x, 2y)$ which are collinear (on top of each other) vectors and a general solution does not exist. We called these vectors as linearly dependent because the vector elements satisfy the linear relation $Q = 2P$ that is, $P - 0.5Q = 0$ holds. More generally a test for linear dependence is as follows.

Linear Dependence Definition: Vectors $p_1, p_2, \ldots p_n$ are linearly dependent if there exist weights $w_1, \ldots w_n$ (not all zero) satisfying the relation

$$w_1 p_1 + w_2 p_2 + \ldots + w_n p_n = 0. \tag{3.10}$$

If such a relation does not exist, the vectors are said to be linearly independent.

If $p_1, p_2, \ldots p_n$ are linearly dependent then one of these vectors can be written as a linear combination (weighted sum) of remaining vectors.

In our second example, $P = (2, 4), Q = (5, 2)$ the solution exists because the two vectors are not collinear. They do not satisfy the test of equation Eq. (3.10) and therefor they are linearly independent. These examples make it plausible that solution exists only when vectors are linearly independent.

We postpone a discussion of tools for finding the solution weights or determining when the vectors are linearly dependent, since use of matrix algebra is particularly powerful for these tasks. We note in passing that matrix algebra methods here involve inversion of a 2×2 matrix and multiplication of the inverse matrix by the vector $(r_1, r_2) = (99, -3)$. Suffice it to say here that R software readily yields the solution weights $w_1 = -13.1250$ and $w_2 = 31.3125$.

The spanning set of vectors generating all points in an n dimensional space can be any linearly independent set of n vectors. Unit vectors are any vectors having Euclidean length unity. When $n = 2$ unit basis vectors are $E_1 = (1, 0)$ and $E_2 = (0, 1)$. Graphically they are simply at unit values on the horizontal axis and vertical axis, respectively. Verify that E_1 and E_2 are orthogonal since the axes are perpendicular to each other. The angle between them should be 90 degrees by Eq. (3.4). The two axes indeed do have the cosine of the angle between them to be zero radians (implying that they are perpendicular or orthogonal to each other).

Basis of a Space created from a set of vectors: Let $(p_1, p_2, \ldots p_n)$ be a collection of vectors in a spanning space V. Now the collection $(p_1, p_2, \ldots p_n)$ forms a basis of the space V if the vectors span V and if they are linearly independent. A set of n linearly independent vectors of a space V of dimension n given in a definite order is called a basis of the space.

When $n = 3$ unit basis vectors are $E_1 = (1, 0, 0), E_2 = (0, 1, 0)$ and $E_3 = (0, 0, 1)$, each of which contain $n = 3$ elements. In general, the unit basis is sometimes called the 'canonical' basis for $\Re^n$.

3.2.2 *Vector Space Defined*

An n dimensional vector space V is a space spanned by a set of vectors $(p_1, p_2, \ldots p_n)$ such that: (a) the sum of any two elements of V remains an element of V, and (b) a scalar multiple of any element also remains an element of V.

As a practical matter, given that x, y and z are elements of a vector space V and given that a, b are numbers, the classic reference Gantmacher (1959) states seven rules:

(1) $x + y = y + x$
(2) $(x + y) + z = x + (y + z)$
(3) there exists a number 0 in V such that the product $0x = 0$
(4) there exists a number 1 in V such that the product $1x = x$
(5) $a(bx) = (ab)x$
(6) $(a + b)x = ax + bx$
(7) $a(x + y) = ax + ay$.

A hands-on experience with these rules can be obtained in R as follows. We choose simple sequences 'x=1:3', 'y=5:7' and choose 'a=21, b=11.'

```
# R program snippet 3.2.1.1 is next.
x=1:3; y=5:7; a=21; b=11
z=c(7,5,9)
x+y; y+x
(x+y)+z ; x + (y+z)
a*(b*x); (a*b)*x
```

```
(a+b)*x;  a*x+b*x
a*(x+y);  a*x + a*y
```

Note that multiplication must be denoted by a (*) in R commands and the semicolon (;) helps print two items one below the other for easy comparison to verify that they are indeed identical as the theory predicts.

```
> x+y; y+x
[1]   6   8 10
[1]   6   8 10
> (x+y)+z ; x + (y+z)
[1] 13 13 19
[1] 13 13 19
> a*(b*x); (a*b)*x
[1] 231 462 693
[1] 231 462 693
> (a+b)*x; a*x+b*x
[1] 32 64 96
[1] 32 64 96
> a*(x+y); a*x + a*y
[1] 126 168 210
[1] 126 168 210
```

Euclidean space is the most common vector space used in this book. More general vector spaces called Banach and Hilbert spaces are often used in engineering, natural and statistical sciences. The reader is referred to the URL:

(http://en.wikipedia.org/wiki/Vector_space) for further details.

3.3   Sum of Vectors in Vector Spaces

The vector in a broad sense represents any quantity possessing a direction. In solid analytic geometry, the vector is defined in a narrow sense as a directed line segment from an initial point such as the origin to a terminal point. The x and y coordinates of a point may be stated in a vector as $A = (x, y) = (2, 3)$, and plotted in R as indicated earlier. Now we consider sum of two vectors in the following snippet.

R program snippet **3.3.1** is next.

```
x=2; y=3  # need to set up the plot size
plot(x,y, typ="1", xlim=c(-1,3), ylim=c(-1,4),
main="Sum of Two Vectors From the Origin", ylab="y")
text(0,0,"O") #show the origin with letter O
arrows(0,0,2,3); text(1.5,3,"A")
arrows(0,0,0.5,0.5); text(0.7,0.4, "B")
arrows(0,0,2.5,3.5, lty=2) #line type 2 for dashed line
text(2.5,3.7,"A+B")
arrows(0,0,-0.5,-0.5); text(-0.7,-0.7, "- B", lty=1)
```

The vector A is depicted as the line OA. We also plot the short vector $B = (0.5, 0.5)$ as the line OB.

Applying the "parallelogram law" for sum of two vectors, one finds the resultant vector $A + B$ graphically as the dashed line $O, A + B$ in Fig. 3.3. It is obtained more easily by simply adding the respective coordinates.

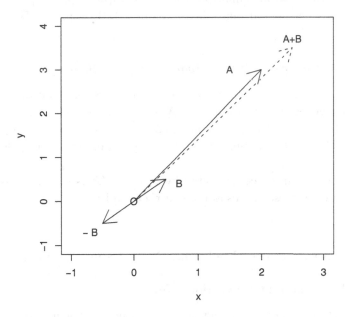

Fig. 3.2   Vectors $A = (2, 3)$ and $B = (0.5, 0.5)$ and their sum as the resultant vector $A + B = (2.5, 3.5)$.

The Fig. 3.3 also depicts the negative vector $-B$ by the line $O-B$ with the arrow suggesting the South West direction.

3.3.1 *Laws of Vector Algebra*

The seven rules from Gantmacher (1959) defining a vector space are related to similar rules for vector algebra. A single magnitude without any 'direction' is called a scalar. When we multiply a vector by a scalar, each element gets multiplied by that scalar. For example, $A = (2, 3)$, $p = 2$, $pA = (4, 6)$. We have already encountered scalar multiplication when we defined $-B$ above, since changing the signs to change the direction amounts to multiplication by the scalar -1.

Let A, B, C denote vectors and p, q denote scalars.

(1) Addition is commutative : A + B = B + A
(2) Addition is associative : A + (B + C) = (A + B) + C
(3) Scalar multiplication is Associative : p(qA) = pq(A) = q(pA)
(4) Distributive two scalar multiplication : (p + q)A = pA + qA
(5) Distributive one scalar multiplication : p(A + B) = pA + pB

3.3.2 *Column Space, Range Space and Null Space*

If X is a $T \times p$ matrix, (e.g. data matrix in regression problems). The vector space generated by columns of X is also called the range space of X. The subspace of all vectors Z in Euclidean space of dimension p such that $XZ = 0$ is called the null space of X, and its dimension is called the nullity.

Note that column space generated by X and $X'X$ is the same. In fact the dimensions of these spaces also equal their ranks.

3.4 Transformations of Euclidean Plane Using Matrices

3.4.1 *Shrinkage and Expansion Maps*

Recall that the vector OQ of Figure 3.1 has Euclidean length 5.3851648. Suppose we want to shrink it to equal the length 4.4721360 of the vector OP. If we denote the coordinates of new shrunken vector as $\tilde{x}, \tilde{y}$. These are related to the coordinates x, y of OQ by the following relation:

$$\begin{bmatrix} \tilde{x} \\ \tilde{y} \end{bmatrix} = \begin{bmatrix} k & 0 \\ 0 & k \end{bmatrix} \begin{bmatrix} x \\ y \end{bmatrix}, \tag{3.11}$$

where we have a shrinkage transformation yielding a shorter vector OQ' when we choose $k = (4.4721360/5.3851648) < 1$. If, on the other hand, we wanted an expansion transformation, we will choose $k > 1$.

3.4.2 *Rotation Map*

Recalling Figure 3.1 and Sec. 3.4.1 consider two vectors OP and OQ' of the same length. Now we want to rotate or map the OQ' vector onto the OP vector. Let the coordinates of OQ' be denoted by $(\tilde{x}, \tilde{y})$ and coordinates of its image OP be denoted as (x', y').

The map from OQ' to OP is a (clockwise) rotation of the vector by an angle denoted by θ. The rotation transformation can be represented by a matrix involving sines and cosines of θ, which equals 48.3664611 degrees according to the output of the first snippet of Sec. 3.1.2. It is possible to write new coordinates of OP by the following map from OQ'.

$$\begin{bmatrix} x' \\ y' \end{bmatrix} = \begin{bmatrix} cos\theta & -sin\theta \\ sin\theta & cos\theta \end{bmatrix} \begin{bmatrix} \tilde{x} \\ \tilde{y} \end{bmatrix}, \tag{3.12}$$

where the matrix involving sines and cosines is called rotation matrix .

The matrix can be viewed as an 'operator' working on the original vector. We shall see that the operation is simply post-multiplication of the rotation matrix by the column vector $(\tilde{x}, \tilde{y})$. It yields the desired column vector (x', y') representing coordinates of the rotated vector. The rotation matrix as an operator represents rotation of the plane about the origin through the angle θ.

3.4.3 *Reflexion Maps*

The mirror images of vectors keeping the y axis fixed and x axis fixed are given by reflection matrices A_y, A_x defined as:

$$A_y = \begin{bmatrix} -1 & 0 \\ 0 & 1 \end{bmatrix}, and\ A_x \begin{bmatrix} 1 & 0 \\ 0 & -1 \end{bmatrix}. \tag{3.13}$$

Reflexions with respect to the line $y = x$ and $y = -x$ are given by reflection matrices $A_{y=x}, A_{y=-x}$ defined as:

$$A_{y=x} = \begin{bmatrix} 0 & 1 \\ 1 & 0 \end{bmatrix}, and\ A_{y=-x} \begin{bmatrix} 0 & -1 \\ -1 & 0 \end{bmatrix}. \tag{3.14}$$

Exercise: Use R functions for matrix multiplication '%*%' and 'det' for determinant to verify (a) that product of reflection matrices is a rotation matrix, and (b) that the determinant of each of the reflection matrices in Eq. (3.13) and Eq. (3.14) is -1.

3.4.4 *Shifting the Origin or Translation Map*

Since the origin is at $(0,0)$ the coordinates of a vector are (x,y). If the origin is shifted to (x^o, y^o), the coordinates become $(x + x^o, y + y^o)$. The image of this transformation is expressed in matrix form by:

$$
\begin{bmatrix} x' \\ y' \\ 1 \end{bmatrix} = \begin{bmatrix} 1 & 0 & x^o \\ 0 & 1 & y^o \\ 0 & 0 & 1 \end{bmatrix} \begin{bmatrix} x \\ y \\ 1 \end{bmatrix},
\tag{3.15}
$$

where the middle matrix is called translation matrix which shifts the origin.

3.4.5 *Matrix to Compute Deviations from the Mean*

Consider a simple $n \times 1$ vector x. The mean is $\bar{x} = \Sigma_{i=1}^n x_i$. The vector y of deviations from the mean has the elements $(x_1 - \bar{x}, x_2 - \bar{x}, \ldots, x_n - \bar{x})$. It is possible to define a matrix

$$
A = I_n - (1/n)\iota\iota' = (1/n)J_n,
\tag{3.16}
$$

where ι denotes a column vector of ones and J_n denotes an $n \times n$ matrix of all ones. This is checked in the following snippet for $n = 4$

```
# R program snippet 3.4.5.1 is next.
set.seed(435); x=sample(1:100,4); t(x)
iota=rep(1,4)
I4=diag(4)
iota.tiota= iota %*% t(iota)
A=I4-(iota.tiota)/4; A
t(A%*%x)#y=Ax, deviations from mean
sum(A%*%x)#deviations add up to zero.
```

The output of the snippet 3.4.5.1 below shows that for $n = 4$ the pre-multiplication by A defined by Eq. (3.16) does give deviations from the mean verified by noting that their sum equals zero.

```
> set.seed(435); x=sample(1:100,4); t(x)
     [,1] [,2] [,3] [,4]
[1,]   80   52   26   83
> iota=rep(1,4)
> I4=diag(4)
> iota.tiota= iota %*% t(iota)
> A=I4-(iota.tiota)/4; A
        [,1]   [,2]   [,3]   [,4]
[1,]   0.75 -0.25 -0.25 -0.25
[2,]  -0.25  0.75 -0.25 -0.25
[3,]  -0.25 -0.25  0.75 -0.25
[4,]  -0.25 -0.25 -0.25  0.75
> t(A%*%x)#y=Ax, deviations from mean
        [,1]  [,2]    [,3]  [,4]
[1,]  19.75 -8.25 -34.25 22.75
> sum(A%*%x)#deviations add up to zero.
[1] 0
```

3.4.6 *Projection in Euclidean Space*

Matrix algebra has established that A is a projection matrix if and only if (iff) it is symmetric idempotent matrix. Hence, let us use the term 'projection matrix' to be synonymous with 'symmetric idempotent matrices,' to be discussed later in Sec. 9.6. Since projection is an intuitive transformation of the Euclidean space, some comments are included here, even though readers unfamiliar with some matrix algebra will not fully understand them.

Consider a linear equation $y = Ax$, where y, x are $n \times 1$ vectors and A is an $n \times n$ matrix. We can call the equation as a *projection* of x into a vector space spanned by the columns of A, provided there exists a symmetric idempotent matrix S such that $y = Sx$. One way to find the needed S is: $S = AA^-$, where A^- denotes the 'generalized inverse' of A discussed in Chapter 16.

Chapter 4

Matrix Basics and R Software

We have discussed vectors and vector spaces in earlier chapters including Chap. 3. Since this book intends to give the reader a 'hands on' experience using R, we point out subtle differences in the concept of a matrix in R versus the concept in mathematics. A matrix can be formed by combining vectors, by 'binding' rows of vectors. The R software has the function 'rbind' for this purpose. Of course, such binding requires that the vectors have identical number of elements. A matrix can also be formed by binding the columns of vectors using the 'cbind' function of R. R will insist that the binding makes sense. Unless the count of the number of items bound together is comparable, R does not allow binding to take place.

4.1 Matrix Notation

In mathematics it is customary to write a typical element of a set in curly braces. In matrix algebra, a matrix is a set of numbers. Hence many authors denote the set of all the elements in a matrix by writing out the typical element inside curly braces.

If $\{a_{ij}\}$ with $(i = 1, 2, \cdots, m)$ and $(j = 1, 2, \cdots, n)$ is a set of precisely ordered real numbers arranged in m rows and n columns, we have an m by n matrix $A = \{a_{ij}\}$ of dimension $m \times n$. It is customary that the order of subscripts follow a specific pattern. The first subscript is for the row number and the second subscript is for the column number. For example, a_{21} refers to the element in the second row and first column.

Now we illustrate a matrix having $m = 2$ rows and $n = 3$ columns. Define a 2×3 matrix A as:

$$A = \begin{bmatrix} a_{11} & a_{12} & a_{13} \\ a_{21} & a_{22} & a_{23} \end{bmatrix} = \begin{bmatrix} 15 & -12 & 53 \\ 25 & 22 & -2 \end{bmatrix} \tag{4.1}$$

A typical matrix in mathematics illustrated by Eq. (4.1) has two dimensions, m, n. If we select one row of A we have a submatrix. Note that in mathematical theory any matrix where $m = 1$ or $n = 1$ is a vector. Hence a vector must be a special case of a matrix. Even a simple number or a scalar is also a special case of a matrix with $m = 1$ and $n = 1$. The designers of R do not allow potential logical conflicts by allowing us to call scalars or vectors as matrices. The R software insists that all matrix objects in R must have an attribute called 'dim' for its two dimensions. The reader must understand this distinction for a seamless hands-on experience using R.

It is possible to construct the matrix illustrated in Eq. (4.1) by using one of three methods: (i) binding rows using 'rbind' (ii) binding columns using 'cbind', or (iii) direct construction using the 'matrix' function of R. We now illustrate the first two methods.

```
# R program snippet 4.1.1 using rbind /cbind.
r1=c(15, -12, 53); r2=c(25, 22, -2)
Ar=rbind(r1,r2); Ar; attributes(Ar)
c1=c(15,25); c2=c(-12, 22); c3=c(53,-2)
dimnames(Ar)[[2]]=c("revenue","cost","profit")
Ar
Ac=cbind(c1,c2,c3); Ac; attributes(Ac)
\vspace*{5pt}
```

The snippet 4.1.1 uses the basic 'c' function to combine numbers into R objects 'r1' and 'r2' for the two rows and 'c1' to 'c3' for the three columns. It also shows how to 'bind' the rows and columns together by the R functions 'rbind' and 'cbind' respectively. The snippet defines an R object 'Ar' and the R function 'attributes' reports its attributes.

In the following output '$dim' reports the dimension of the matrix to be 2×3. Note that since 'Ar' is comprised of rows named 'r1' and 'r2', R assigns the rows of 'Ar' matrix these names. R calls them '$dimnames[[1]]' or the names of dimension 1 (row dimension) being the objects 'r1' and 'r2'. Since columns of 'Ar' do not have any particular names, R assigns sequence numbers [,1] to [,3] as column names. The R command 'attributes(Ar)' used in the snippet 4.1.1 applies the R function 'attributes' to the matrix object 'Ar' and leads in the output below to the conclusion that '$dim-names[[2]]' is NULL, implying that the dimension 2 (for columns) has no names. R allows us to assign any names to columns if we wish to do so. Indeed, we call them 'revenue', 'cost' and 'profit' by using the command 'dimnames(Ar)[[2]]=c("revenue","cost","profit")'.

```
> Ar=rbind(r1,r2); Ar; attributes(Ar)
   [,1] [,2] [,3]
r1   15  -12   53
r2   25   22   -2
$dim
[1] 2 3
$dimnames
$dimnames[[1]]
[1] "r1" "r2"
$dimnames[[2]]
NULL
> c1=c(15,25); c2=c(-12, 22); c3=c(53,-2)
> dimnames(Ar)[[2]]=c("revenue","cost","profit")
> Ar
   revenue cost profit
r1      15  -12     53
r2      25   22     -2
> Ac=cbind(c1,c2,c3); Ac; attributes(Ac)
     c1  c2 c3
[1,] 15 -12 53
[2,] 25  22 -2
$dim
[1] 2 3
$dimnames
$dimnames[[1]]
NULL
$dimnames[[2]]
[1] "c1" "c2" "c3"
```

Note that the matrix 'Ac' formed by binding the columns is the same as the one shown in Eq. (4.1). The 'attributes' function of R shows that the objects Ac and Ar are indeed matrix objects with two dimensions. To repeat, R assigns 'names' to rows and columns from what it knows to be the ingredients forming the matrix. Note also that dimnames[[1]]

has row names and dimnames[[2]] has column names in the standard order whereby rows come first and then come columns. The names are character vectors and must be defined the R function 'c' with elements in quotation marks and separated by a comma. This is illustrated by the R command 'dimnames(Ar)[[2]]=c("revenue","cost","profit")' of the snippet 4.1.1.

4.1.1 *Square Matrix*

A square matrix is a special kind of matrix with as many rows as are columns. In the above example if we had three rows instead of two, it would have been a square matrix. A square matrix with m rows and n columns satisfies: $m = n$.

Similar to many authors let us implicitly assume that all vectors are column vectors, unless stated otherwise. Indeed this is the convention used in R software for vectors and also when matrices are created out of long vectors.

4.2   Matrices Involving Complex Numbers

A matrix is a function on a set of integers $i \in [1, \ldots, m], j \in [1, \ldots, n]$, with values in the field of real numbers $\Re$ or complex numbers $\mathscr{C}$. The corresponding totality of matrices is denoted as $M_{m,n}(\Re)$ or $M_{m,n}(\mathscr{C})$. We indicate a specific $m \times n$ matrix as $A \in M_{m,n}(\mathscr{C})$ generally only when complex numbers are involved. If nothing is specified it is understood a typical matrix $A \in M_{m,n}(\Re)$. If A is a $m \times m$ square matrix it is understood that $A \in M_m(\Re)$. We explicitly write $A \in M_m(\mathscr{C})$ when complex numbers are admitted.

A complex conjugate of a scalar quantity $a = x + iy$ is $\bar{a} = x - iy$ where $i = \sqrt{(-1)}$, $a\bar{a} = \parallel a \parallel^2$ and $a + \bar{a}$ are always real. By analogy the complex conjugate $\bar{A}$ of a matrix A replaces all entries by their respective complex conjugates.

$$\text{If} A = \begin{bmatrix} 1 + i & i \\ 2 - i & 4 \end{bmatrix}, \text{ then } \bar{A} = \begin{bmatrix} 1 - i & -i \\ 2 + i & 4 \end{bmatrix} \tag{4.2}$$

A theorem in matrix algebra states that A is real if and only if $\bar{A} = A$. Also $A + \bar{A}$ is always real.

A^* denotes the complex adjoint of A. The (i,j)-th entry of A^* is the complex conjugate of the (j,i)-th entry of A. We shall see in Sec. 4.5 that such interchanging of rows and columns is called the 'transpose' in matrix algebra. That is, complex adjoint is a conjugate transpose.

Complex numbers are commonly used in study of cyclical phenomena or spectral analysis of time series. A simple example of complex adjoint or conjugate transpose denoted by the superscript (*) is given next.

$$\text{If} \quad A = \begin{bmatrix} 0 & i \\ i & 0 \end{bmatrix}, \text{ then } A^* = (\bar{A})' = \begin{bmatrix} 0 & -i \\ -i & 0 \end{bmatrix}, \tag{4.3}$$

where $\bar{A}$ denotes the complex conjugate.

4.3   Sum or Difference of Matrices

Denote the matrix elements as: $A = \{a_{ij}\}$, $B = \{b_{ij}\}$, and $C = \{c_{ij}\}$. We can write $c_{ij} = a_{ij} + b_{ij}$ only when both A and B are of exactly the same dimension. Then, their sum is obtained by simply adding terms precisely matched for their row and column numbers.

For example, note that $A + B$ is a matrix of all zeros when A is given by Eq. (4.1) and B is defined to have same elements as A but opposite signs. We define

$$B = \begin{bmatrix} b_{11} & b_{12} & b_{13} \\ b_{21} & b_{22} & b_{23} \end{bmatrix} = \begin{bmatrix} -15 & 12 & -53 \\ -25 & -22 & 2 \end{bmatrix} \tag{4.4}$$

Having seen how to construct a matrix in R by binding rows or columns, we turn to the third method. One can construct a matrix by giving a long column-wise sequence all numbers in it to the R function 'matrix' while providing the dimensions. For example, the matrix A is created in R by first defining A as a long catalog of all numbers in it using the 'c' function of R and then using the command 'matrix(A,nrow=2,ncol=3)' to define the row dimension, nrow=2, and column dimension as ncol=3. See the following snippet of R commands.

```
# R program snippet 4.3.1 defining a matrix.
A = c(15,25,-12,22,53,-2) #column-wise list all numbers
A=matrix(A,nrow=2,ncol=3)#nrow in matrix function means
#number of rows,
#ncol inside matrix function of r means number of columns.
B=c(-15,-25,12,-22,-53,2)  #all elements have opposite signs
B=matrix(B,2,3)  #matrix assumes two numbers as nrow and ncol
C=A+B
A;B;C
```

The R output of the snippet 4.3.1 is abridged for brevity. Note that matrix rows are reported by R with rows names [1,], [2,], etc. with the second part after the comma reserved for column number empty. The columns are reported with headings [,1], [,2], etc with the first part before the comma empty. These are conventions for extracting sub matrices.

```
A
        [,1] [,2] [,3]
[1,]  15  -12   53
[2,]    25   22   -2


B
        [,1] [,2] [,3]
[1,]  -15   12  -53
[2,]   -25  -22    2


C
        [,1] [,2] [,3]
[1,]     0    0    0
[2,]     0    0    0
```

We have defined B by changing the signs of all elements of A. The snippet verifies that $C = A + B = 0$ holds true.

R allows complete and easy access to individual elements and specific rows and columns of a matrix object by using a notation involving **bracketed numbers**. For example, the snippet 4.3.2 using the input of the snippet 4.3.1 shows how we can access (print to screen) the entire second column of A, th entire first row of B and finally the [2,2] element along the second row and second column of C.

```
# R program snippet 4.3.2 subsetting.
A[,2]
B[1,]
C[2,2]
```

Once we access the column vector, it is stored as a 'list' or 'vector' object by R and reported in a compact fashion to the screen.

```
> A[,2]
[1] -12   22
> B[1,]
```

```
[1] -15  12 -53
> C[2,2]
[1] 0
```

We emphasize that in R the brackets [,] play a special intuitively obvious role in extracting individual elements or arrays of elements, where the convention always row numbers come first and column numbers come last, separated by a comma. The parentheses in R reserved for evaluating 'functions' should not be confused with brackets.

4.4 Matrix Multiplication

When we multiply two numbers a, b (scalars or 1×1 matrices) we simply multiply them as $a * b$ which always equals $b * a$. Not so if we are multiplying matrices! This section will show that the order of multiplication does matter for matrices A, B. That is, matrix multiplication AB is not commutative $AB \neq BA$. In fact, it is possible that one or more of the matrix multiplications may not be well defined (conformable).

Recall the curly braces notation and let $A = \{a_{ij}\}$ be $r_a \times c_a$ matrix with $i = 1, 2, \cdots, r_a$ rows and $j = 1, 2, \cdots, c_a$ columns. Similarly let $B = \{b_{k\ell}\}$ matrix have $k = 1, 2, \cdots, r_b$ rows, and $\ell = 1, 2, \cdots, c_b$ columns. Let us understand the notion of matrix conformability before discussing matrix multiplication. If A is $r_a \times c_a$ matrix, and B is is $r_b \times c_b$ matrix, then the row column multiplication of AB is conformable (well defined) only if

$$\text{Number of columns of } A = c_a = r_b = \text{Number of rows of } B. \qquad (4.5)$$

An easy way to remember this is that in any "row-column" multiplication, we must have equality of the numbers of "column-rows".

Row-column Multiplication: Now the matrix multiplication of A and B gives the product matrix $C = \{c_{i\ell}\}$ which is of dimension $r_a \times c_b$, and is obtained by "row-column" multiplication as follows.

$$c_{i\ell} = \sum_{t=1}^{c_a} a_{it} b_{t\ell}, i = 1, 2, \cdots, r_a, \ell = 1, 2, \cdots, c_b \qquad (4.6)$$

Note that a_{it} in the summation represent the elements of i-th row as t ranges over all r_a columns of A. Also, the $b_{t\ell}$ in the summation represent the elements of ℓ-th column again as t ranges over all c_b columns of B. The summation is well-defined only if the range of t is precisely matched. That is, only if $c_a = r_b$, or Eq. (4.5) holds true.

Examples of conformable matrix multiplications:

(1) $A_{2\times3}\, B_{3\times5}$,
(2) $A_{2\times3}\, B_{3\times5}$,
(3) $A_{2\times3}\, B_{3\times1}$,
(4) $A_{5\times4}\, B_{4\times50}$,
(5) $A_{5\times1}\, B_{1\times5}$ (called outer product),
(6) $A_{1\times5}\, B_{5\times1}$.

where we have denoted the matrix dimensions by self-explanatory subscripts. Note that all of them satisfy the condition in Eq. (4.5).

In R, row-column multiplication of equation Eq. (4.6) via the summation notation is obtained rather easily. Matrix multiplication of A with B is achieved by simply writing A %*% B, where the special notation '%*%' means row-column multiplication and is obviously distinct from a simple multiplication of two numbers (scalars) denoted by '*' in R. For example '2*3=6.'

Let us first consider a multiplication of two 2×2 matrices which are obviously conformable with a simple example so the reader can fully understand the row-column multiplication by making simple mental multiplications.

R program snippet **4.4.1** Defines matrix multiplication.
```
P=matrix(1:4,2,2); P
Q=matrix(11:14,2,2); Q
PQ=P%*%Q; PQ
```

In the row-column multiplication we multiply elements from (1,3) based on the first row of P by the corresponding elements (11,12) based on the first column of Q. Thus the element in the [1,1] location of PQ denoted in the following R output of the snippet 4.4.1 by 'P%*%Q' is $(1 \times 11 + 3 \times 12) = 11 + 36 = 47$. Similarly the element in the [1,2] location of PQ is $(1 \times 13 + 3 \times 14) = 13 + 42 = 55$.

```
> P=matrix(1:4,2,2); P
     [,1] [,2]
[1,]   1    3
[2,]   2    4
> Q=matrix(11:14,2,2); Q
      [,1] [,2]
[1,]   11   13
[2,]   12   14
> PQ=P%*%Q; PQ
```

```
      [,1] [,2]
[1,]   47   55
[2,]   70   82
```

If matrix multiplication in R is attempted without making sure that Eq. (4.5) holds true, R responds with an error message as illustrated below.

R program snippet **4.4.2** Faulty matrix multiplication.
```
A = c(15,25,-12,22,53,-2)
A=matrix(A,2,3)
B=c(-15,-25,12,-22,-53,2)   #all elements have opposite signs
B=matrix(B,2,3)
A%*%B  #this is the multiplication command
```

The R output of the faulty snippet 4.4.2 below indicates and 'Error' condition.

```
Error in A %*% B : non-conformable arguments
```

In the following snippet we consider $A_{2\times3}$ with $r_a = 2, c_a = 3$ as before and a new $B_{3\times3}$ having $r_b = 3 = c_b$. Now row-column multiplication satisfies the column-rows equality requirement $c_a = r_b = 3$, from Eq. (4.5).

R program snippet **4.4.2** valid matrix multiplication.
```
A = c(15,25,-12,22,53,-2)
A=matrix(A,2,3)
B=matrix(1:9,3,3);B #new B
A%*%B  #this is the multiplication command
```

We have chosen simple elements of B for easy mental calculations.

```
> B=matrix(1:9,3,3);B #new B
      [,1] [,2] [,3]
[1,]    1    4    7
[2,]    2    5    8
[3,]    3    6    9
> A%*%B  #this is the multiplication command
      [,1] [,2] [,3]
[1,]  150  318  486
[2,]   63  198  333
```

The reader is asked to create several simple matrices on paper and in R by using the snippet 4.4.2 as a template. Then the reader should do matrix multiplications on paper to gain a clear understanding of row-column multiplications. Comparing paper calculations with the R output will provide a hands-on experience with instant feedback.

4.5   Transpose of a Matrix and Symmetric Matrices

The transpose of a matrix $A = \{a_{i,j}\}$ is another matrix created from A matrix by interchanging rows and columns. It is usually denoted by the transpose (dash or prime) symbol A' or by the superscript A^T. Thus $A' = \{a_{j,i}\}$.

R allows easy computation of the transpose by the function 't' for transpose.

```
# R program snippet 4.5.1 for transpose of a matrix.
A = c(15,25,-12,22,53,-2)
A=matrix(A,2,3); A
t(A) #transpose of A
```

In response to above commands R yields the following output.

```
> A=matrix(A,2,3); A
     [,1] [,2] [,3]
[1,]   15  -12   53
[2,]   25   22   -2
> t(A) #transpose of A
     [,1] [,2]
[1,]   15   25
[2,]  -12   22
[3,]   53   -2
```

Note that the first row of A is made the first column in t(A) or A'

4.5.1   *Reflexive Transpose*

The operation of transpose is reflexive meaning that transpose of transpose gives back the original matrix: $(A')' = A$.

4.5.2 *Transpose of a Sum or Difference of Two Matrices*

If two matrices A and B are conformable in the sense that their sum and difference are well defined, then we have: $(A+B)' = A'+B'$, and $(A-B)' = A' - B'$.

4.5.3 *Transpose of a Product of Two or More Matrices*

A rule for the transpose of a product of two matrices AB (not necessarily square matrices) or three matrices ABC is:

$$(AB)' = B'A', \quad (ABC)' = C'B'A', \tag{4.7}$$

where the order of multiplication reverses.

Now we use R to check this rule. We define A, B and AB, compute the transpose of AB and check if it equals the right hand side of Eq. (4.7)

```
# R program snippet 4.5.3.1 Transpose AB
#R.snippet checking equality involving transposes
A = c(15,25,-12,22,53,-2)
A=matrix(A,2,3); #as before
B=matrix(1:9,3,3);B
LHS=t(A%*%B) #left side of rule
RHS=t(B) %*% t(A)#right side of rule
LHS-RHS #should be a matrix of zeros
```

The output of snippet 4.5.3.1 below shows that left side does equal right side, since the difference is a 3×2 matrix of zeros.

```
> B=matrix(1:9,3,3);B
     [,1] [,2] [,3]
[1,]    1    4    7
[2,]    2    5    8
[3,]    3    6    9
> LHS=t(A%*%B) #left side of rule
> RHS=t(B) %*% t(A)#right side of rule
> LHS-RHS #should be a matrix of zeros
     [,1] [,2]
[1,]    0    0
[2,]    0    0
[3,]    0    0
```

4.5.4 *Symmetric Matrix*

Definition of a Symmetric Matrix: If the matrix equals its own transpose, that is $A = A'$, then it is said to be symmetric.

Symmetric matrices have identical elements below and above the main diagonal.

A matrix algebra theorem states that: *The product of any matrix and its transpose yields a symmetric matrix.* That is, the matrix multiplication $Q = AA'$ always yields a symmetric matrix Q. We can verify this fact analytically by checking whether the transpose $Q' = (AA')'$ equals Q. This is indeed true since by Section 4.5.3 we have

$$Q' = (AA')' \tag{4.8}$$
$$= (A')'A' \tag{4.9}$$
$$= AA' = Q,$$

where the last equation is written by using the result in Section 4.5.1.

The simplest way to construct examples of symmetric matrices in R is to use Eq. (4.8). Let us randomly create matrix A from columns of data for variables x, y, z, by binding their columns. Next we can use 't(A) %*% A' as a matrix multiplication to create a symmetric matrix. However, the following snippet uses the 'crossprod' function of R which accomplishes the same matrix multiplication, but is usually slightly faster.

```
# R program snippet 4.5.4.1 is next.
set.seed(452); x=rnorm(20)
y=rnorm(20);z=rnorm(20) #random vectors
sym22=crossprod(cbind(x,y));sym22
t(sym22) #transpose of 2 by 2
sym33=crossprod(cbind(x,y,z));sym33
t(sym33)#transpose of 3 by 3
```

Note that sym22 illustrates a 2×2 symmetric matrix and sym33 illustrates a 3×3 symmetric matrix. Observe how the elements across the main diagonal from top left to bottom right are equal in both cases in the following output of the snippet 4.5.4.1.

```
> sym22=crossprod(cbind(x,y));sym22
         x          y
x 25.911577  6.025813
y  6.025813 18.944357
```

```
> t(sym22) #transpose of 2 by 2
          x         y
x 25.911577  6.025813
y  6.025813 18.944357
> sym33=crossprod(cbind(x,y,z));sym33
           x          y          z
x 25.9115765  6.025813 -0.8822119
y  6.0258134 18.944357 -2.4824467
z -0.8822119 -2.482447 16.8965957
> t(sym33)#transpose of 3 by 3
           x          y          z
x 25.9115765  6.025813 -0.8822119
y  6.0258134 18.944357 -2.4824467
z -0.8822119 -2.482447 16.8965957
```

Note that we can visually verify that Section 4.5.1 result that transpose of symmetric matrices equals the original matrix in these examples.

4.5.5 *Skew-symmetric Matrix*

Definition of a Skew-symmetric Matrix: If the matrix equals the negative of its own transpose, then it is said to be skew-symmetric. $A = -A'$

For example, a 3×3 skew-symmetric matrix is defined as:

$$A = \begin{bmatrix} 0 & a_{12} & a_{13} \\ -a_{12} & 0 & a_{23} \\ -a_{13} & -a_{23} & 0 \end{bmatrix} \qquad (4.10)$$

Only square matrices can be symmetric or skew-symmetric. A theorem in matrix algebra states that: *The difference of any matrix and its transpose is a skew-symmetric matrix.*

We can analytically verify the theorem as follows. By Section 4.5.2 we have

$$\begin{aligned} (A - A')' &= (A') - A \\ &= -(A - A'), \end{aligned} \qquad (4.11)$$

where the last equation is written by using the result in Section 4.5.1. It is skew-symmetric because it equals the negative of its own transpose per definition of such matrices given above.

Any square matrix A can be written as a sum of two parts: where the symmetric part is $(1/2)(A + A')$ and the anti-symmetric part is $(1/2)(A - A')$.

The R package 'Matrix,' Bates and Maechler (2010), allows easy computation of these two parts as shown in the following snippet. The user has to first download the package from the Internet (using the 'packages' menu of the R GUI or graphical user interface) on his computer and then bring it into the current memory of R by using the 'library' command.

\# R program snippet **4.5.5.1** skew-symmetric matrix.

```
library(Matrix) #bring package into current R memory
m <- Matrix(1:4, 2,2); m
symmpart(m)#symmetric part
skewpart(m)#skew-symmetric part
```

The R function 'sympart' gets the symmetric part and 'skewpart' gets the skew-symmetric part in the following output of snippet 4.5.5.1 .

```
> m <- Matrix(1:4, 2,2); m
2 x 2 Matrix of class "dgeMatrix"
     [,1] [,2]
[1,]    1    3
[2,]    2    4
> symmpart(m)#symmetric part
2 x 2 Matrix of class "dsyMatrix"
     [,1] [,2]
[1,]  1.0  2.5
[2,]  2.5  4.0
> skewpart(m)#skew-symmetric part
2 x 2 Matrix of class "dgeMatrix"
     [,1] [,2]
[1,]  0.0  0.5
[2,] -0.5  0.0
```

Note that the top right corner element 3=2.5+.5. It is easy to check mentally that the two parts add up to the original matrix 'm' with elements 1 to 4.

4.5.6   *Inner and Outer Products of Matrices*

Note that R assumes that all vectors are column vectors. A product of two column vectors D and E by usual row-column multiplication does not exist, since the requirement Eq. (4.5) fails to hold. This is verified in the following snippet.

\# R program snippet **4.5.6.1** Faulty inner product.
```
D=1:5 #assumed to be column vector
E=11:15 #column vector
D%*%E
```

Since the snippet 4.5.6.1 is faulty, its output is 'Error'.

```
Error in D %*% E : non-conformable arguments
```

We can make the two $n \times 1$ column vectors D and E conformable for vector multiplication of D and E in two ways: (1) Replace the first vector D by a row vector, or (2) replace the second vector E by a row vector. We can replace the first vector D as a $1 \times n$ row vector by transposing it as D', and then write the matrix multiplication as: $D'E$ which is a 1×1 scalar. This is called the inner or cross product of two vectors in Eq. (3.2).

Alternatively, we can replace the second vector E as a $1 \times n$ row vector by transposing it as E', and then write the matrix multiplication as: DE' which is a $n \times n$, possibly large square matrix. This is called the outer product matrix.

\# R program snippet **4.5.6.1** Valid inner product.
```
D=1:5 #assumed to be column vector
E=11:15 #column vector
t(D)%*%E #inner product
D%*%t(E) #outer product
```

Note that inner product or cross product is a scalar ($=205$) whereas the outer product is a matrix. Note that the element in the [2,3] position of the outer product matrix is a product of D[2]=2 with E[3]=13 or 26.

```
> t(D)%*%E #inner product
     [,1]
[1,]  205
> D%*%t(E) #outer product
     [,1] [,2] [,3] [,4] [,5]
```

```
[1,]    11    12    13    14    15
[2,]    22    24    26    28    30
[3,]    33    36    39    42    45
[4,]    44    48    52    56    60
[5,]    55    60    65    70    75
```

Matrix Multiplication is Not Commutative:

In general we do *not* expect the multiplication of A and B matrices to satisfy $AB = BA$. In fact BA may not even exist even when AB is well defined, and vice versa. In the (rare) special case when $AB = BA$ holds true, we say that matrices A and B commute.

4.6   Multiplication of a Matrix by a Scalar

If h is a scalar element and A is a matrix, then $hA = Ah = C$ defined by $C = \{c_{ij}\}$, where $c_{ij} = h\,a_{ij}$.

In R, the simple multiplication operator (*) should be used when a scalar is involved. Row column multiplication operator (%*%) does not work when multiplying a matrix by a scalar. The order of multiplication does not matter: $h * A$ and $A * h$ give exactly the same answer.

```
# R program snippet 4.6.1 is next.
#R.snippet Scalar times matrix in R
A = c(15,25,-12,22,53,-2)
A=matrix(A,2,3); A
h=9
A%*%h   #Incorrect Row column multiplication
A*h; h*A   #both give same correct answer

> A=matrix(A,2,3); A
     [,1] [,2] [,3]
[1,]   15  -12   53
[2,]   25   22   -2
> h=9
> A%*%h   #Incorrect Row column multiplication
Error in A %*% h : non-conformable arguments
> A*h; h*A   #both give same correct answer
     [,1] [,2] [,3]
[1,]  135 -108  477
```

```
[2,]   225   198   -18
     [,1]  [,2]  [,3]
[1,]   135  -108   477
[2,]   225   198   -18
```

The output of snippet 4.6.1 shows that 'hA' and 'Ah' give identical answers.

4.7 Multiplication of a Matrix by a Vector

Let A be $r_a \times c_a$ as above, and $x = (x_1, x_2, \cdots x_{r_a})$ be a $1 \times r_a$ row vector. Now the multiplication Ax is not defined. To see this, simply treat multiplication of a matrix by a vector as a product of two matrices above. On the other hand, if x is a column vector $c_a \times 1$, then Ax is a well defined $r_a \times 1$ column vector.

```
# R program snippet 4.7.1 R function matrix.
A = c(15,25,-12,22,53,-2) #start with a vector
A=matrix(A,2,3); A
x=c(9,11,2)
# the 3 numbers in the object x needs to be rearranged
#use matrix function to define a row vector in R
x=matrix(x,1,3)#since x is row vector Ax fails
A%*%x  # Use Row-column multiplication in R
x=matrix(x,3,1) #define a COLUMN vector in R
A%*%x  #Row column multiplication in R
```

The output of snippet 4.7.1 shows that the matrix multiplication 'Ax' fails when 'x' is a row vector, but computes when 'x' is a column vector.

```
> A=matrix(A,2,3); A
     [,1]  [,2]  [,3]
[1,]   15   -12    53
[2,]   25    22    -2
> x=c(9,11,2)
> # the 3 numbers in the object x needs to be rearranged
> #use matrix function to define a row vector in R
> x=matrix(x,1,3)#since x is row vector Ax fails
> A%*%x  # Use Row-column multiplication in R
Error in A %*% x : non-conformable arguments
> x=matrix(x,3,1) #define a COLUMN vector in R
```

```
> A%*%x  #Row column multiplication in R
     [,1]
[1,]  109
[2,]  463
```

Multiplication of a matrix by a vector is a fundamental operation in matrix algebra. For example, it is useful for writing a system of equations in matrix notation. Recall the following system of two equations from Eq. (3.8).

The system

$$p_{11}\,w_1 + p_{12}\,w_2 = r_1, \tag{4.12}$$

$$p_{21}\,w_1 + p_{22}\,w_2 = r_2 \tag{4.13}$$

is written in matrix notation as $Pw = r$, where $P = \{p_{ij}\}, i = 1, 2$, and $j = 1, 2$. The matrix P is 2×2 and the column vectors w and r are both 2×1. The w vector is defined from (w_1, w_2) as the unknown vector of weights. The r vector is defined from (r_1, r_2), known right hand values.

It is important to note that one can use matrix algebra to formally express a general systems of equations. We must follow the rules for row-column multiplication even when we use symbols instead of numbers as matrix and/or vector elements.

4.8　Further Rules for Sum and Product of Matrices

Since we need to define the identity matrix for some of these rules, let us begin this section with its definition and how to get it in R.

Diagonal and Identity Matrices Defined:

The identity matrix I_n is an $n \times n$ matrix with elements $I_n = \{\delta_{ij}\}$, based on the delta function, that is, where $\delta_{ij} = 1$ when $i = j$ and $\delta_{ij} = 0$ otherwise.

The identity matrix is a diagonal matrix. Verify that all nonzero elements of I_n are only along its main diagonal, hence it is called a diagonal matrix.

In R the $n \times n$ identity matrix is produced by the function 'diag(n)'. For $n = 3$ we have the identity matrix as the following R output.

```
> diag(3) #identity matrix n=3
     [,1] [,2] [,3]
[1,]    1    0    0
```

```
[2,]    0    1    0
[3,]    0    0    1
```

In scalar arithmetic raising 1 to any integer power gives 1 again. Similarly, note that (row-column) matrix multiplication of Identity matrix with itself (any number of times) gives the identity matrix back. That is, the identity matrix is idempotent (See Sec. 9.6 for details)

```
> diag(2)%*%diag(2)
     [,1] [,2]
[1,]    1    0
[2,]    0    1
```

If a vector of (nonzero) values is given, a diagonal matrix based on them in R is produced by the 'diag' function.

```
> diag(7:8)
     [,1] [,2]
[1,]    7    0
[2,]    0    8
```

Finally we are ready to list the rules for matrix sums and products. We assume that additions and multiplications are conformable.

(1) **Associativity.** For sums we have:
$A + B + C = A + (B + C) = (A + B) + C$;
and for matrix multiplications: $ABC = A(BC) = (AB)C$.

(2) **Distributivity.** $A(B + C) = AB + AC$;
$(B + C)A = BA + CA$.

(3) **Getting back the original matrix in matrix additions**:
Null Matrix: Define the null matrix as $O_{m,n}$, an $m \times n$ matrix of all zeros. Now, adding the null matrix as in: $A + O_{m,n} = A$, gets back the original A matrix of dimension $m \times n$.

(4) **Getting back the original matrix in matrix multiplications**:
Now matrix multiplication will get back the original matrix if one multiplies by the identity matrix: $AI_n = A$.

The last rule has practical relevance. Note that it allows cancellation and simplification of complicated matrix expressions. Given a number n, if we multiply it by its reciprocal $1/n$, we have unity: $n * (1/n) = 1$. In matrix algebra the reciprocal of matrix A is called its 'inverse' and denoted as A^{-1}, by the superscript (-1). We shall see in Sec. 8.2 that the 'inverse' is defined so that $AA^{-1} = I$, is the identity matrix. Hence, the trick in simplification usually involves making some expressions equal to the identity matrix or zero matrix in a string of matrix expressions. In usual arithmetic we do not bother to write out multiplications by 1 or 0. In matrix algebra also it is not necessary to write out multiplication by $I, 0$ matrices, hence the simplification.

The following example of simplification may be skipped by readers who are new to matrix inverses. Assume that one wants to verify that the 'hat matrix' defined as $H = X(X'X)^{-1}X'$ is idempotent in the sense that $H^2 = H$. If we write a long expression

$$H^2 = HH = X(X'X)^{-1}[X'X(X'X)^{-1}]X' = H, \qquad (4.14)$$

we note that $(X'X)^{-1}(X'X)$ appearing in the middle of the long expression is simply the identity matrix and can be ignored. Hence Eq. (4.14) simplifies to $H = X(X'X)^{-1}X'$. In other word we have used simplifications to prove that $HH = H$. If $X = \iota$ is a column of ones, then note that $XX' = J_n$ is an n-square matrix of all ones, encountered earlier in Sec. 3.4.5. The hat matrix for this X becomes $(1/n)J_n$ in Eq. (3.16).

4.9   Elementary Matrix Transformations

Using matrix multiplications and certain rearrangements of the identity matrix we can study three types of elementary matrix transformations. We consider a nonzero scalar k, $m \times n$ matrix A and $I_m = m \times m$ identity matrix. We illustrate them below in the R snippet 4.9.1

E1) Row or Column Interchange: If we wish to interchange row $i_1 \leq m$ with row $i_2 \leq m$, first make the interchange operation on I_m matrix and call the resulting matrix as I_{row}. Now the matrix multiplication $I_{row}A$ accomplishes the desired interchange.

E2) Scalar times Row or Column: If we wish to multiply row i or column j with k, first make the scalar multiplication operation on I_m matrix and call the resulting matrix as I_{scalar}. Now the matrix multiplication $I_{scalar}A$ accomplishes the desired scalar multiplication by a nonzero scalar.

E3) Add to any row (column) a scalar times corresponding elements of another row (column): The addition is done element by element. For example, we want to add to row $i_1 \leq m$ element by element k times corresponding element of row $i_2 \leq m$, first make the operations on I_m matrix and call the resulting matrix as I_{E3}. Now the matrix multiplication $I_{E3}A$ accomplishes the desired interchange.

In the following illustrative snippet using R we have chosen $m = 4, n = 3, k = 25, i_1 = 2, i_2 = 4$.

Some known results regarding E1 to E3 transformations are: (1) Any nonsingular matrix can be written as a product of matrices of type E1 to E3. (2) The rank and norm of a matrix are not changed by E1 to E3. (3) Two matrices A, B of the same dimension are **equivalent** if B can be obtained from pre or post multiplying A by a finite number of elementary matrix transformations.

```
# R program snippet 4.9.1 Elementary Matrix Transforms.
A=matrix(1:12,4,3); A
ide=diag(4);ide #the identity matrix
ideRow=ide #initialize
ideRow[2,]=ide[4,] #replace row 2 by row 4
ideRow[4,]=ide[2,]#replace row 4 by row 2
ideRow#view transformed identity matrix
ideRow %*% A #verify E1 interchange
k=25
idescalar=ide #initialize
idescalar[2,2]=k #second row has k
idescalar %*% A #Verify E2 transf with k
ideE3=ide #initialize
i4=k*ide[4,]
ideE3[2,]=ide[2,]+i4
ideE3# view E3 operator matrix
ideE3 %*% A #Verify E3 on A

> A=matrix(1:12,4,3); A
     [,1] [,2] [,3]
[1,]    1    5    9
[2,]    2    6   10
[3,]    3    7   11
[4,]    4    8   12
> ide=diag(4);ide #the identity matrix
     [,1] [,2] [,3] [,4]
[1,]    1    0    0    0
```

```
[2,]    0    1    0    0
[3,]    0    0    1    0
[4,]    0    0    0    1
> ideRow=ide #initialize
> ideRow[2,]=ide[4,] #replace row 2 by row 4
> ideRow[4,]=ide[2,]#replace row 4 by row 2
> ideRow#view transformed identity matrix
     [,1] [,2] [,3] [,4]
[1,]    1    0    0    0
[2,]    0    0    0    1
[3,]    0    0    1    0
[4,]    0    1    0    0
> ideRow %*% A #verify E1 interchange
     [,1] [,2] [,3]
[1,]    1    5    9
[2,]    4    8   12
[3,]    3    7   11
[4,]    2    6   10
> k=25
> idescalar=ide #initialize
> idescalar[2,2]=k #second row has k
> idescalar %*% A #Verify E2 transf with k
     [,1] [,2] [,3]
[1,]    1    5    9
[2,]   50  150  250
[3,]    3    7   11
[4,]    4    8   12
> ideE3=ide #initialize
> i4=k*ide[4,]
> ideE3[2,]=ide[2,]+i4
> ideE3# view E3 operator matrix
     [,1] [,2] [,3] [,4]
[1,]    1    0    0    0
[2,]    0    1    0   25
[3,]    0    0    1    0
[4,]    0    0    0    1
> ideE3 %*% A #Verify E3 on A
     [,1] [,2] [,3]
[1,]    1    5    9
```

```
[2,]  102  206  310
[3,]    3    7   11
[4,]    4    8   12
```

The snippet 4.9.1 shows the realization of elementary matrix transformations for rows. Similar operation for columns are easily created by starting with I_n identity matrix with dimension equal to the number of columns in A and using post-multiplication of A by suitably defined transformed identity matrix. For example adding $k = 10$ times third column to the first column is illustrated by the code in the following snippet.

R program snippet **4.9.2** Elementary Column Operations.

```
#assume previous snippet is in memory of R
k=10
ide3=diag(3); ide3E3=ide3 #initialize
i4=k*ide3[,3] #k times third column
ide3E3[,1]=ide3[,1]+i4#add to first column
ide3E3# view E3 operator matrix
A%*%ide3E3 #Verify E3 on A for columns
```

We remind the reader that when elementary row (column) matrix transformations are involved, one uses pre (post) multiplication by suitably transformed identity matrix.

```
> k=10   #use k=10 for easy visualization
> ide3=diag(3); ide3E3=ide3 #initialize
> i4=k*ide3[,3] #k times third column
> ide3E3[,1]=ide3[,1]+i4#add to first column
> ide3E3# view E3 operator matrix
     [,1] [,2] [,3]
[1,]    1    0    0
[2,]    0    1    0
[3,]   10    0    1
> A%*%ide3E3 #Verify E3 on A for columns
     [,1] [,2] [,3]
[1,]   91    5    9
[2,]  102    6   10
[3,]  113    7   11
[4,]  124    8   12
```

The snippets show that elementary operations are easily implemented in R for any size matrices.

4.9.1 *Row Echelon Form*

Consider a 4×4 matrix with the following pattern:

$$A = \begin{bmatrix} 1 & 0 & a_{13} & a_{14} \\ 0 & 1 & a_{23} & a_{24} \\ 0 & 0 & 1 & a_{34} \\ 0 & 0 & 0 & 0 \end{bmatrix}, \tag{4.15}$$

where all nonzero rows are above the all zero row, and where the leading number (the first nonzero number when going from the left, also called the pivot) of a nonzero row is always strictly to the right of the leading coefficient of the row above it. If Eq. (4.15) has $a_{13} = 0 = a_{23}$, so that we have a clean 3×3 identity matrix in its top left corner, then it is called 'reduced' row-echelon form.

It is easy to imagine general row-echelon matrices where there can be several rows with all zero elements and where A is not a square matrix. Every nonzero matrix can be reduced via elementary matrix transformations discussed in Sec. 4.9 to a unique 'reduced' row-echelon form. There are two elimination methods available for such reductions.

Gaussian elimination is an algorithm for solving systems of linear equations, finding the rank (see Sec. 7.2) of a matrix, and calculating the inverse (see Sec. 8.2) of a nonsingular (see Sec. 6.4) matrix. Gaussian elimination uses elementary row operations to reduce a matrix to row echelon form.

The Gauss-Jordan elimination is an extension of Gaussian elimination which further reduces the result to 'reduced' row-echelon form.

Both Gaussian and Gauss-Jordan elimination methods are outdated tools in today's computing environment, since matrix inversion and finding the rank of a matrix can be directly accomplished, without involving any cumbersome row-echelon forms.

4.10 LU Decomposition

A square matrix A can be written as a product of two square matrices of the same order, lower-triangular matrix L and upper-triangular matrix U:

$$A = LU, \tag{4.16}$$

where lower-triangular matrix has all its elements above the main diagonal zero and upper-triangular matrix has all its elements below the main diagonal zero. The R package 'Matrix' computes the decomposition as shown in the snippet 4.10.1.

```
# R program snippet 4.10.1 LU decomposition.
library(Matrix)#download package Matrix
set.seed(99);A=matrix(sample(1:100,9),3)
A #print original matrix
str(luA <- lu(A))#define luA
eluA=expand(luA)#expand luA
eluA#has 3 parts P=permutation, L and U
eluA$L %*% eluA$U #need not equal A
#needs pre-multiplication by eluA$P
A2 <- with(eluA, P %*% L %*% U);A2
```

Sparse matrix is a matrix whose entries are mostly zeros. Since the 'Matrix' package has highly sophisticated R functions designed to do economical storage of sparse matrices, it needs the 'expand' function and also uses a permutation matrix P for economical computations of the LU algorithm. The net effect of it is that one needs to use $A = PLU$ instead of Eq. (4.16), which is not entirely satisfactory. If we change the original matrix to PA then the software does yield $PA = LU$ as in Eq. (4.16).

The notation 'dgeMatrix' in the ouput from the above R commands given below refers to dense general matrix (defined as not sparse), 'pMatrix' refers to numeric matrices in packed storage, and 'dtrMatrix' refers to triangular dense matrix in non-packed storage.

```
> set.seed(99);A=matrix(sample(1:100,9),3)
> A #print original matrix
     [,1] [,2] [,3]
[1,]   59   97   64
[2,]   12   52   28
[3,]   68   92   33
> str(luA <- lu(A))#define luA
Formal class 'denseLU' [package "Matrix"] with 3 slots
  ..@ x   : num [1:9] 68 0.176 0.868 92 35.765 ...
  ..@ perm: int [1:3] 3 2 3
  ..@ Dim : int [1:2] 3 3
```

```
> eluA=expand(luA)#expand luA
> eluA#has 3 parts P=permutation, L and U
$L
3 x 3 Matrix of class "dtrMatrix" (unitriangular)
        [,1]        [,2]        [,3]
[1,]  1.0000000                .            .
[2,]  0.1764706  1.0000000                .
[3,]  0.8676471  0.4802632  1.0000000
$U
3 x 3 Matrix of class "dtrMatrix"
      [,1]      [,2]      [,3]
[1,]  68.00000  92.00000  33.00000
[2,]        .  35.76471  22.17647
[3,]        .          .  24.71711
$P
3 x 3 sparse Matrix of class "pMatrix"
[1,] . . |
[2,] . | .
[3,] | . .
> eluA$L %*% eluA$U #need not equal A
3 x 3 Matrix of class "dgeMatrix"
     [,1] [,2] [,3]
[1,]   68   92   33
[2,]   12   52   28
[3,]   59   97   64
> A2 <- with(eluA, P %*% L %*% U) #
> A2
3 x 3 Matrix of class "dgeMatrix"
     [,1] [,2] [,3]
[1,]   59   97   64
[2,]   12   52   28
[3,]   68   92   33
```

The above snippet shows that the matrix A2 obtained after LU decomposition of A and matrix multiplication is the same as the original matrix A. The LU decomposition is not difficult in R even for large matrices.

Chapter 5

Decision Applications: Payoff Matrix

Now that we have defined vectors and matrices in some detail, we want to dispel the notion that matrix algebra is for mathematicians. We show how algebraic tools can be used in everyday decision making. This chapter uses some principles from elementary decision theory. We introduce various criteria used in decision analysis around a carefully created payoff matrix which summarizes the problem faced by the decision maker. The very construction of a payoff matrix is a very useful start in making rational decisions in day-to-day life of an individual or a business. The principles remain relevant for very large multinational corporations.

5.1 Payoff Matrix and Tools for Practical Decisions

One is often faced with practical decisions involving limited knowledge and uncertainty. In elementary statistics texts it is recommended that we create an $n \times m$ payoff matrix usually having human decisions $D_1, D_2, \ldots D_n$ listed along the matrix rows and external reality outside the control of the decision maker (U for uncontrolled) listed along the columns $U_1, U_2, \ldots, U_m$.

For example, the human decider can build a 'D_1 =large, D_2 =medium, or D_3 =small' factory. The decision problem amounts to choosing among these three ($n = 3$) alternatives. The external reality beyond the control of the decider or 'uncontrolled by the decider' may offer a 'U_1 =fast growing, U_2 =medium, or U_3 =slow' economic environment with three ($m = 3$) possibilities. The $n \times m$ payoff matrix contains estimates of payoffs (present value of ultimate profits over a decision horizon, net of all costs) for each pair of $U_i, i = 1, .., n$ and $D_j, j = 1, .., m$. Good decisions need good estimates of the payoff matrix.

Now we fill up the nine numbers of the payoff matrix column by column based on prior knowledge. If the economy is fast growing, the profit is forecast to be (100, 70, 20), if the decider builds a large, medium or small factory, respectively. This will be the first column of payoff matrix called U_1 for the first of almighty's options. If the economy is medium growth, the analogous profits are: (65, 70, 45), representing U_2. If the economy is slow growing, the respective profits are (30, 40, 25) along the third column of the payoff matrix. This is set up in the following R snippet.

R program snippet **5.1.1** for Payoff Matrix setup.

```
U1=c(100, 70, 20) #profits for fast growing economy
U2=c(65, 70, 45) #profits for medium economy
U3=c(30, 40, 25) #profits for slow economy
#payoff=matrix(c(U1,U2,U3),nrow=3,ncol=3)
payoff=cbind(U1,U2,U3)
dimnames(payoff)[[1]]=c("D1","D2","D3")
#dimnames(payoff)[[2]]=c("U1","U2","U3")
payoff
```

The snippet 5.1.1 uses c(.,.) function of R to define column entries: U1, U2 and U3. Combining the columns into a long list of numbers is again done with the same c function as: c(U1,U2,U3). The 'matrix' function of R then creates the payoff matrix when we specify the size of the matrix. Note that care is needed in typing the information since R is case sensitive. For example, 'c' and 'matrix' as names of R functions must be lower case. Actually it is easier to construct the payoff matrix by binding together the three columns by the R function 'cbind.' The snippet has commented out the matrix command with a # at the beginning of the command line. The 'dimnames(payoff)[[1]]' assigns row names and 'dimnames(payoff)[[2]]' assigns column names. The names must be character vectors defined with quotation marks.

```
> payoff
    U1 U2 U3
D1 100 65 30
D2  70 70 40
D3  20 45 25
```

5.2 Maximax Solution

The command 'apply(payoff, 2, max)' computes an R object called 'row-max' containing the maximum along each row decision. The number 2 in the command 'apply' refers to the second (column) dimension of the matrix. The maximax solution to the decision problem is overly optimistic assuming that the best uncertain will always happen. This chooses the best column first. Among the maxima of each decision, the maximax principle chooses the largest number. Then within the rows of the best column, choose the highest value. In the above table, the max(rowmax) is 100 along the first row implying that the decision maker should choose the first row decision D_1 to build a large factory.

```
# R program snippet 5.2.1 Maximax decision.
#R.snippet bring in previous snippet
payoff #make sure payoff matrix is present
rowmax=apply(payoff, 2, max);rowmax
max(rowmax) #maximax solution
gr1=which(rowmax == max(rowmax))
gr2=which(payoff[,gr1] == max(rowmax))
#print(c("maximax solution decision",gr2))
maximax=row.names(payoff)[gr2]
print(c("maximax solution", maximax))
```

The above snippet 5.2.1 uses the R function 'which' twice to find 'gr1' the sequence number (location) where the best column is located. The function 'which' also finds 'gr2' as the best row within the chosen column to locate the the maximax solution as the row name of 'payoff' matrix at 'gr2'. This allows us to pinpoint the decision itself in a general fashion.

```
> rowmax=apply(payoff, 2, max);rowmax
 U1  U2  U3
100  70  40
> max(rowmax) #maximax solution
[1] 100
> gr1=which(rowmax == max(rowmax))
> gr2=which(payoff[,gr1] == max(rowmax))
> #print(c("maximax solution decision",gr2))
> maximax=row.names(payoff)[gr2]
> print(c("maximax solution", maximax))
```

```
[1] "maximax solution" "D1"
```

The above output of snippet 5.2.1 shows that the maximax solution is to build a large factory or D_1.

5.3   Maximin Solution

The command 'apply(payoff, 2, min)' computes an R object called 'rowmin' containing the minimum profit associated with each decision. The maximin solution to the decision problem is overly pessimistic assuming that the worst will always happen. This lists the worst performing values for each column first as (20, 45, 25) respectively. Among the minima of each decision, the maximin principle chooses the largest number. This happens to be in the second column at 45. Now our task to find the decision associated with 45 in the second column, which is along the third row implying that the decision maker should choose the third row decision to build a small factory, as can be intuitively predicted from the pessimism of the decision maker.

```
# R program snippet 5.3.1 Maximin solution.
#R.snippet bring in previous snippet
payoff #make sure payoff matrix is present
rowmin=apply(payoff, 2, min);rowmin
max(rowmin) #maximin solution
gr1=which(rowmin == max(rowmin))#decision number
gr2=which(payoff[,gr1] == max(rowmin))
maximin=row.names(payoff)[gr2]
print(c("maximin choice", maximin))
decisions=rbind(maximax, maximin)
dimnames(decisions)[[2]]=c("Decisions by Various Criteria so far")
print(decisions)
```

In the snippet 5.3.1 we have collected two solutions so far, in a matrix object of R called 'decisions'. Note that the column name for this spells out what the matrix will have by using the 'dimnames' function of R. Now the output of above code is:

```
> rowmin=apply(payoff, 2, min);rowmin
U1 U2 U3
20 45 25
```

```
> max(rowmin) #maximin solution
[1] 45
> gr1=which(rowmin == max(rowmin))#decision number
> gr2=which(payoff[,gr1] == max(rowmin))
> maximin=row.names(payoff)[gr2]
> print(c("maximin choice", maximin))
[1] "maximin choice" "D3"
> decisions=rbind(maximax, maximin)
> dimnames(decisions)[[2]]=c("Decisions by Various Criteria so far")
> print(decisions)
        Decisions by Various Criteria so far
maximax "D1"
maximin "D3"
```

For the payoff data in the above table, the last line of the output of snippet 5.3.1 shows the maximin solution at 'max(rowmin)', where profit=45, implying the decision D_2 to build a medium size factory.

5.4 Minimax Regret Solution

Opportunity loss matrix or 'regret' matrix measures along each column what 'might have been' in terms of maximum profit based on the column maximum. This section focuses on decision makers that are known to focus on what they "might have earned," had they decided differently. The following snippet first computes the regret matrix by subtracting each entry of the column from the column maximum. For example, if the economy is strong, the best profit 100 comes from building a large factory. If one builds a medium factory the regret equals $(100 - 70)$, whereas if one builds a small factory the regret equals $(100 - 20)$. Similarly for each column we subtract the current entries from the respective column maximum.

Now review the regret matrix and make sure that it has a zero along the row where the column maximum is present. Next we compute the maximum regret associated with each decision row. This summarizes the maximum regret (worst feeling) produced by each decision. The minimax regret principle *minimizes* the maximum regret (worst feeling) instead of focusing on maximizing the best feeling as in the earlier criteria.

```
# R program snippet 5.4.1 for Minimax Regret.
#Assume that previous snippet is in R memory
```

```
payoff #verify that payoff matrix is in memory
colmax=apply(payoff, 2, max);colmax
U1r=colmax[1]-U1;U1r#col 1 of regret table
U2r=colmax[2]-U2;U2r
U3r=colmax[3]-U3;U3r
#regret=matrix(c(U1r,U2r,U3r),nrow=3,ncol=3)
regret=cbind(U1r,U2r, U3r)
dimnames(regret)[[1]]=c("D1","D2","D3")
regret #print regret table
rrowmax=apply(regret, 1, max);
print(c("maximum regret", rrowmax))
print(c("minimum of maximum regret", min(rrowmax)))
gr1=grep(min(rrowmax),rrowmax)#decision number
minimax.regret=row.names(payoff)[gr1]
decisions=rbind(maximax, maximin, minimax.regret)
dimnames(decisions)[[2]]=c("Decisions by Various Criteria so far")
decisions
```

Note that the R object containing the regret matrix is also called 'regret.'
Verify that it has a zero value at the column column maximum of the payoff
matrix, signifying no regret when maximum profit is achieved.

```
> colmax=apply(payoff, 2, max);colmax
 U1   U2   U3
100   70   40
> U1r=colmax[1]-U1;U1r#col 1 of regret table
[1]   0 30 80
> U2r=colmax[2]-U2;U2r
[1]   5  0 25
> U3r=colmax[3]-U3;U3r
[1] 10  0 15
> #regret=matrix(c(U1r,U2r,U3r),nrow=3,ncol=3)
> regret=cbind(U1r,U2r, U3r)
> dimnames(regret)[[1]]=c("D1","D2","D3")
> regret #print regret table
   U1r U2r U3r
D1   0   5  10
D2  30   0   0
D3  80  25  15
> rrowmax=apply(regret, 1, max);
```

```
> print(c("maximum regret", rrowmax))
                              D1                  D2                  D3
"maximum regret"            "10"                "30"                "80"
> print(c("minimum of maximum regret", min(rrowmax)))
[1] "minimum of maximum regret" "10"
> gr1=grep(min(rrowmax),rrowmax)#decision number
> minimax.regret=row.names(payoff)[gr1]
> decisions=rbind(maximax, maximin, minimax.regret)
> dimnames(decisions)[[2]]=c("Decisions by Various Criteria so far")
> decisions
                Decisions by Various Criteria so far
maximax         "D1"
maximin         "D2"
minimax.regret  "D1"
```

The above computation shows that minimax regret gives the solution of the maximum regret of 10 obtained by building a large factory, that is, choosing decision D_1.

5.5 Digression: Mathematical Expectation from Vector Multiplication

Uncertainty is studied in elementary Statistics by computing mathematical expectation associated with each outcome. This is an important topic and it helps to review it in standard notation in Statistics before proceeding further. Let the outcomes (payoffs) be denoted by $x_1, x_2, \ldots x_m$ and the corresponding probabilities of the occurrence of those outcomes are known to be $p_1, p_2, \ldots p_m$ such that $\Sigma_{i=1}^m p_i = 1$, implying that the list x_i is mutually exclusive and exhaustive. Now mathematical expectation is defined as $E(x) = \Sigma_{i=1}^m x_i p_i$. One way to deal with uncertainty is to compute $E(x)$ associated with each decision. In most cases, it is appropriate to choose the decision yielding maximum expected value $E(x)$. It is generally wise not to enter into gambling activities where the expected value is negative.

Now we review the mathematical expectation in terms of vectors and multiplication of two vectors. If x is a column vector of outcomes and p is a column vector of probabilities, expected value $E(x) = x'p$, where the prime on x denotes a transpose of the vector. One obtains the same answer if we pre-multiply x by the transpose p'. After finishing this digression regarding $E(x)$, we are ready to go back to the main theme of payoff matrices and decisions.

5.6　Maximum Expected Value Principle

This method requires additional data. Each uncertain outcome $U_1, U_2, \ldots U_m$ must have a probability of occurrence of that outcome $p_1, p_2, \ldots p_m$ available to the decider. These probabilities must add up to unity, that is the list of uncertain choices along the columns must be mutually exclusive and exhaustive. For our example let us choose $p_1 = 0.3, p_2 = 0.5, p_3 = 0.2$ which do add up to unity as they should. Snippet 5.6.1 illustrates the factory size choice decision for these probabilities.

```
# R program snippet 5.6.1 for max expected value.
#Assume R memory has payoff and regret matrices
prob=c(0.3, 0.5, 0.2)#specify new data on probabilities
sum(prob)#this should equal unity
expected.value=payoff %*%prob; expected.value
maxev=max(expected.value)
print(c("maximum Expected Value", maxev))
gr1=grep(max(expected.value), expected.value)
max.Exp.Value=row.names(payoff)[gr1]
max.Exp.Value #in terms of decision number
expected.regret=regret%*%prob; expected.regret
miner=min(expected.regret) #magnitude
print(c("minimum Expected Regret", miner))
gr1=grep(min(expected.regret), expected.regret)
min.Exp.Regret=row.names(payoff)[gr1]
min.Exp.Regret #in terms of decision number
decisions=rbind(max.Exp.Value, min.Exp.Regret)
dimnames(decisions)[[2]]=c("Decisions by Expected Value Criteria")
decisions
```

The following R output reports the result from a matrix multiplication of a payoff or regret matrix with the known vector of probabilities.

```
> prob=c(0.3, 0.5, 0.2)#specify new data on probabilities
> sum(prob)#this should equal unity
[1] 1
```

```
> expected.value=payoff %*%prob; expected.value
  [,1]
D1 68.5
D2 64.0
D3 33.5
> maxev=max(expected.value)
> print(c("maximum Expected Value", maxev))
[1] "maximum Expected Value" "68.5"
> gr1=grep(max(expected.value), expected.value)
> max.Exp.Value=row.names(payoff)[gr1]
> max.Exp.Value #in terms of decision number
[1] "D1"
> expected.regret=regret%*%prob; expected.regret
  [,1]
D1  4.5
D2  9.0
D3 39.5
> miner=min(expected.regret) #magnitude
> print(c("minimum Expected Regret", miner))
[1] "minimum Expected Regret" "4.5"
> gr1=grep(min(expected.regret), expected.regret)
> min.Exp.Regret=row.names(payoff)[gr1]
> min.Exp.Regret #in terms of decision number
[1] "D1"
> decisions=rbind(max.Exp.Value, min.Exp.Regret)
> dimnames(decisions)[[2]]=c("Decisions by Expected Value Criteria")
> decisions
                Decisions by Expected Value Criteria
max.Exp.Value   "D1"
min.Exp.Regret  "D1"
```

Note that building a large factory (decision row 1) maximizes the expected value and also minimizes expected regret for the data at hand.

Cost of Uncertainty The cost of uncertainty is computed as follows. If we had no uncertainty and already knew what the external reality is going to be, we will choose the decision associated with the maximum from payoff in each column.

For our example, the maximum payoffs in the three columns are: (100, 70, 40). Using the corresponding probabilities prob=c(0.3, 0.5, 0.2) we compute the expected value as 't(prob)%*%colmax' command gives 73. The maximum expected value gives 68.5. Hence the cost of uncertainty (not knowing exactly what external reality will be) is (73-68.5)=4.5 for our example.

\# R program snippet **5.6.2** for Cost of Uncertainty.
```
#Assume R memory has payoff and regret matrices
colmax #make sure these are col.max of payoff
max.certain=t(prob)%*%colmax
cost.of.uncertainty=max.certain-max(expected.value)
cost.of.uncertainty
```

The following R output reports the result from a matrix multiplication of a prob with 'colmax' and goes on to compute the cost of uncertainty.

```
> colmax #make sure these are col.max of payoff
 U1   U2   U3
100   70   40
> max.certain=t(prob)%*%colmax
> cost.of.uncertainty=max.certain-max(expected.value)
> cost.of.uncertainty
      [,1]
[1,]   4.5
```

These computations can be extended to large problems with a large number n of decision choices and a large number m of uncertainties in columns of the payoff table. Use of R in these extensions removes the drudgery from all calculations and allows the decision maker to focus on getting most detailed and reliable payoff table.

5.7   General R Function 'payoff.all' for Decisions

It is not difficult to write a general R function which will take payoff matrix and vector of column probabilities as input, make all computations and report all decisions.

\# R program snippet **5.7.1** for all decisions together.
```
payoff.all=function(payoff,prob){
n=nrow(payoff)
m=ncol(payoff)
```

```
m1=length(prob)
if(sum(prob)!= 1)  "Error: sum input prob should be 1"
if(m1!=m) "Error: Number of prob not equal to No of payoff columns"
#naming rows and columns
namesrow=paste("D",1:n,sep="")
namescol=paste("U",1:m,sep="")
dimnames(payoff)[[1]]=namesrow
dimnames(payoff)[[2]]=namescol
print(payoff) #new row column names
rowmax=apply(payoff, 2, max)
print("maximums in each column")
print(rowmax)
gr1=which(rowmax == max(rowmax))
gr2=which(payoff[,gr1] == max(rowmax))
print(c("maximax solution column number",gr1))
#max(rowmax) #maximax solution
#gr1=grep(max(rowmax),rowmax);gr1#decision number
maximax=row.names(payoff)[gr2]
print(c("maximax choice", maximax))
#now row minimums for each column
rowmin=apply(payoff, 2, min)
print("minimums in each column")
print(rowmin)
print(max(rowmin)) #maximin solution
gr1=which(rowmin == max(rowmin))#decision number
gr2=which(payoff[,gr1] == max(rowmin))
maximin=row.names(payoff)[gr2]
print(c("maximin choice", maximin))
#### Now payoff
payoff #verify that payoff matrix is in memory
colmax=apply(payoff, 2, max);colmax
mar=matrix(rep(colmax,n),n,m,byrow=T)
regret=mar-payoff
dimnames(regret)[[2]]=namescol
dimnames(regret)[[1]]=namesrow
print(" Regret table")
print(regret)
rrowmax=apply(regret, 1, max);
print(c("maximum regret", rrowmax))
```

```
print(c("minimum of maximum regret", min(rrowmax)))
gr1=which(rrowmax == min(rrowmax))#decision number
minimax.regret=row.names(payoff)[gr1]
decisions=rbind(maximax, maximin, minimax.regret)
print("Decisions by Deterministic Criteria")
dimnames(decisions)[[2]]=c("Decisions by Deterministic Criteria")
print(decisions)
####  Now expected value
expected.value=payoff %*%prob;
print("expected values for each decision")
print(expected.value)
maxev=max(expected.value)
print(c("maximum Expected Value", maxev))
#gr1=grep(max(expected.value), expected.value)
gr1=which(expected.value == max(expected.value))
max.Exp.Value=row.names(payoff)[gr1]
print(max.Exp.Value) #in terms of decision number
expected.regret=regret%*%prob;
print("expected regrets for each decision")
print(expected.regret)
miner=min(expected.regret) #magnitude
print(c("minimum Expected Regret", miner))
gr1=which(expected.regret == min(expected.regret))
min.Exp.Regret=row.names(payoff)[gr1]
min.Exp.Regret #in terms of decision number
decisions2=rbind(max.Exp.Value, min.Exp.Regret)
dimnames(decisions2)[[2]]=
c("Decisions by Expected Value Type Criteria")
#print("Decisions by Expected Value Type Criteria")
print(decisions2)
#now compute cost of uncertainty
colmax #make sure these are col.max of payoff
max.certain=t(prob)%*%colmax
cost.of.uncertainty=max.certain-max(expected.value)
list(decisions=decisions, decisions2=decisions2,
cost.of.uncertainty=cost.of.uncertainty)
}
#example set.seed(99);pary1=round(runif(40, min=10, max=50),0)
#payoff.all(matrix(pary1,4,10), prob=rep(.1,10))
```

The reader should try a random example with a 4 by 10 matrix and use the 'payoff.all' function in the fashion indicated along the last two rows of the above snippet.

5.8 Payoff Matrix in Job Search

This section illustrates construction of complicated payoff matrices from job search data involving a lower salary during the early probation time and a higher salary after the employer accepts the employee. We assume a five year time horizon.

A student has 3 job offers A, B and C with eventual annual salaries of 30K, 33K and 35K respectively. Job A comes with a probation period of 1 year with 90% of eventual salary, B with probation of 2 years with 80% of the eventual salary and C with probation of 2 years with probation salary 80% of the eventual salary. The total job contract is for 5 years. There are 3 states of nature depending on the unknown reaction of employer and co-workers defined as (i) the student does the job very well, (ii) moderately well, or (iii) poorly. If the candidate does poorly, he or she is sure to not get any extension beyond the probation period. In the absence of an extension, the candidate will have to work in a job similar to a Burger joint for only $20K. If the candidate performs moderately, he or she will have a job for first 3 years and then will end up with only 20K job for the remaining period out of five years. Construct a payoff table and find various solutions for various choice criteria discussed above.

Ultimate salary (payoff) has 3 components:

(1) Salary during probation years (pay1)
(2) Salary after probation years before getting fired (pay2)
(3) Salary after getting fired at the main job and working at the burger joint.

The payoff table can and should have all three components. column 1) If student performs well, payoff.well=pay1+pay2 column 2) If student performs moderately well, he will survive three years on the job and must work at the burger joint for remaining 2 years. pay3 earnings on the job for 3 yrs pay4 earnings for last 2 years payoff.moderate=pay1+pay3+pay4 column 3) If student performs poorly, he gets burger job beyond probation. pay5=(years beyond probation)*(20K the burger salary) payoff.poor=pay1+pay5 Let the values for the 3 jobs A, B and C be lists in R created by using the c(.,.,.) as a sequence of 3 numbers.

```
# R program snippet 5.8.1 for job search.
salary=c(30,33,35) #at first choice jobs
burger.salary =c(20,20,20)#after being fired
prop.salary.dur.probation=c(0.90,0.80,0.80)
probation.yrs=c(1,2,2)#job A has shorter probation
yrs.after.probation=5-probation.yrs # 4 3 3 for A,B and C
remain.yr1=c(2,1,1)
pay1=salary*prop.salary.dur.probation*probation.yrs
print(pay1)
#[1] 27.0 52.8 56.0
pay2=yrs.after.probation*salary
pay2
#[1] 120  99 105
payoff.well=pay1+pay2
yrs.after.probationTill3=3-probation.yrs
pay3=yrs.after.probationTill3*salary
pay4=2*burger.salary #for last two years
payoff.moderate=pay1+pay3+pay4
pay5=yrs.after.probation*burger.salary
payoff.poor=pay1+pay5
payoff.tab=cbind(payoff.well,
payoff.moderate, payoff.poor)
dimnames(payoff.tab)[[1]]=c("A","B","C")
payoff.tab #after naming rows with jobs
```

```
> payoff.tab
  payoff.well payoff.moderate payoff.poor
A       147.0           127.0       107.0
B       151.8           125.8       112.8
C       161.0           131.0       116.0
```

One can apply the general function in Section 5.7 to the above data by the command 'payoff.all(payoff.tab, prob=c(0.2, 0.5, 0.3))'. The abridged output is given below.

```
$decisions
                Decisions by Deterministic Criteria
maximax         "D3"
```

```
maximin          "D1"
minimax.regret "D3"
```

```
$decisions2
               Decisions by Expected Value Type Criteria
max.Exp.Value   "D3"
min.Exp.Regret "D3"
```

It appears that by and large, our analysis suggests choosing decision D_3 or job C, since it is favored by four out of five criteria.

Chapter 6

Determinant and Singularity of a Square Matrix

Mathematicians discovered determinants long before they defined matrices. We have discussed matrix basics in the previous chapter Chap. 4. In the context of a quadratic equation Eq. (2.8) and Eq. (2.9) we note that zero determinant is crucial. A detailed discussion of matrix inversion and characteristic roots comes in following chapters, but remains intimately related to the ideas in this chapter.

Let the matrix A be a square matrix $A = (a_{ij})$ with i and $j = 1, 2, \cdots, n$. The determinant is a scalar number. When $n = 2$ we have a simple calculation of the determinant as follows:

$$det(A) = \begin{vmatrix} a_{11} & a_{12} \\ a_{21} & a_{22} \end{vmatrix} = a_{11}a_{22} - a_{21}a_{22}, \tag{6.1}$$

where one multiplies the elements along the diagonal (going from top left corner to bottom right corner) and subtracts a similar multiplication of diagonals going from bottom to top.

In R the function function 'matrix' creates a square matrix if the argument list neglects to specify a second dimension. For example 'A=matrix(10:13,2)' creates a 2×2 square matrix from the four numbers: (10,11,12,13). In R the function 'det' gives the determinant as shown below. For the following example, the det(A) is -2 and is also obtained by cross multiplication and subtraction by 10*13-11*12.

```
#R.snippet
A=matrix(10:13,2)  #creates a square matrix from 10,11,12,13
A #view  the square matrix
det(A) #compute the determinant
10*13-11*12
```

99

R OUTPUT is as follows:

```
> A #view  the square matrix
      [,1] [,2]
[1,]   10   12
[2,]   11   13
> det(A) #compute the determinant
[1] -2
> 10*13-11*12
[1] -2
```

Note that 'det(A)' is -2, and '$10*13 - 11*12$' is also -2.

When dimension of A is $n = 3$, we can find its determinant by expanding it in terms of three 2×2 determinants and elements of the first row of A as follows:

$$det(A) = \begin{vmatrix} a_{11} & a_{12} & a_{13} \\ a_{21} & a_{22} & a_{23} \\ a_{31} & a_{32} & a_{33} \end{vmatrix} = a_{11}|A_{11}| - a_{12}|A_{12}| + a_{13}|A_{13}|, \qquad (6.2)$$

where the upper case A with subscripts is a new notation: A_{ij} which denotes a submatrix of A formed by erasing the i-th row and the j-th column from the original A. For example, deleting the first row and third column of A we have:

$$A_{13} = \begin{vmatrix} a_{21} & a_{22} \\ a_{31} & a_{32} \end{vmatrix} \qquad (6.3)$$

This is also called a minor, $|A_{ij}|$ and denotes the determinant of the 2×2 matrix.

One can readily verify the expansion of $|A|$ upon substituting for determinants involved in the three minors:

$$|A| = a_{11}a_{22}a_{33} - a_{11}a_{23}a_{32}c + a_{12}a_{23}a_{31} - a_{12}a_{21}a_{33} + a_{13}a_{21}a_{32} - a_{13}a_{22}a_{31}. \qquad (6.4)$$

An interesting and practical trick to compute a 3×3 determinant is to carefully write its numbers as a 3×5 grid with the two extra columns created by simply repeating the first two columns. Note that this expanded matrix has three diagonals top to bottom and three diagonals bottom to top.

Now by analogy with the 2×2 matrix multiply out and add the elements along the three diagonals (going from top left corner to bottom right corner) and subtract a similar multiplication of three diagonals going from bottom to top. One can easily verify that this 'easy' method yields exactly the same six terms as the ones listed above. For example,

$$det \begin{vmatrix} a & d & g \\ b & e & h \\ c & f & j \end{vmatrix}$$

is expanded into a 3×5 set up by simply repeating the first two columns as:

$$\begin{pmatrix} a & d & g & a & d \\ b & e & h & b & e \\ c & f & j & c & f \end{pmatrix}$$

Now we multiply the diagonals going from top left corner to bottom right corner and subtract a similar multiplication of three diagonals going from bottom to top. This gives the solution as $aej + dhc + gbf - ceg - fha - jbd$.

6.1   Cofactor of a Matrix

Signed determinants of minors A_{ij} are called cofactors and are denoted here as:

$$c_{ij} = (-1)^{i+j} |A_{ij}| \tag{6.5}$$

In general, for any n one can expand the determinant of A, denoted by $|A|$ or $det(A)$ in terms of cofactors as

$$det(A) = \sum_{j=1}^{n} a_{ij} c_{ij}, \tag{6.6}$$

where the summation is over j for a fixed i. That is, the right hand side of Eq. (6.6) can be expressed in n different ways. It works for any one row from out of $i = 1, 2, \cdots n$.

Alternatively, we can expand as:

$$det(A) = \sum_{i=1}^{n} a_{ij} c_{ij}, \tag{6.7}$$

where the summation is over i for a fixed j. That is, for any one column from out of $j = 1, 2, \cdots n$.

An alternative elegant definition of a determinant which does not use cofactors at all is

$$det(A) = \Sigma_{\sigma} sgn(\sigma) \Pi_{i=1}^{n} a_{i,\sigma(i)} \tag{6.8}$$

where σ runs over all permutations of integers $1, 2, \cdots$, n and the function sgn(σ) is sometimes called signature function.

The signature function becomes 1 or -1 depending on whether a permutation is odd or even. For example, if n= 2, the even permutation of $\{1,2\}$ is $\{1,2\}$, whereby $\sigma(1) = 1$ and $\sigma(2) = 2$. Now the odd permutation of $\{1,2\}$ is $\{2,1\}$ having $\sigma(1) = 2$ and $\sigma(2) = 1$. Given a permutation, one counts the number of steps needed to get to it from the natural sequence $(1, 2, \ldots, n)$. If one needs even (odd) number of steps, it is said to be an even (odd) permuation.

Exercise: Verify that the determinant of a simple 2×2 matrix A defined by Eq. (6.8) remains $a_{11}a_{22} - a_{12}a_{21}$.

The R package 'e1071,' Dimitriadou *et al.* (2010), offers some useful tools for finding the permutations with known alternate signs starting with $+ - + - + - \cdots$. These can be used for understanding the notion of computing the determinant from permutations of the matrix elements.

```
#R.snippet R tools to generate lists of permutations
library(e1071)
p3=permutations(3);p3
pa=paste(1:3, t(p3)) #need transpose of p3
print(matrix(pa, ncol=3, byrow=T), quote=F)
```

The output from above commands directly gives the subscripts needed for the determinant.

```
     [,1] [,2] [,3]
[1,] 1 1  2 2  3 3
[2,] 1 2  2 1  3 3
[3,] 1 2  2 3  3 1
[4,] 1 1  2 3  3 2
[5,] 1 3  2 1  3 2
[6,] 1 3  2 2  3 1
```

It is easy to check that the list of subscripts in Eq. (6.4) starting with $a_{11}a_{22}a_{33} - a_{11}a_{23}a_{32}c + a_{12}a_{23}a_{31}$, etc. matches with the rows of R output above, except that instead of being in order 1 to 6 we have in the order $(1,4,3,2,5,6)$ for the row labels.

The definition of a determinant based on permutations is generally not used for numerical computations. It is mostly of theoretical interest. The R function 'det' is readily available for numerical work.

6.2   Properties of Determinants

Following properties of determinants are useful, and may be readily verified.
We show limited numerical verification in R snippets.

1] $det(A) = det(A')$, where A' is the transpose of A.

```
#R.snippet det(A) equals det of transpose of A
A=matrix(10:13,2)   #create 2 by 2 matrix from 10,11,12,13
A; det(A) # gives -2 as the answer
# t(A) means transpose of A in R
det(t(A))  # still gives -2 as the answer
```

2] $det(AB) = det(A)det(B)$ for the determinant of a product of two
matrices. Unfortunately, for the determinant of a sum of two matrices
there is no simple equality or inequality: $det(A + B) \neq det(A) + det(B)$.

```
#R.snippet det(AB) equals det(A) times det(B)
A=matrix(10:13,2)
B=matrix(c(3,5,8,9), 2); B
det(A %*% B)  #this is 26
det(B); det(A)
det(A)*det(B) #this is also 26 as claimed by theory
```

3] If B is obtained from A by interchanging a pair of rows (or columns),
then det(B)$= -det(A)$.

In R, the function 'rbind' binds rows into a matrix. Second row of A is
denoted in R by A[2,] and first row of A is given by A[1,]. Interchanging
the rows of A amounts to making the second row as the first and the first
row as the second. Hence the new B matrix is created by the command
'B=rbind(A[2,],A[1,]).' For this example, $det(B)$ is seen to be $+2$, which is
just a change of sign of $det(A)$ which was found to be -2 above.

```
#R.snippet determinant changes sign with row interchange
A=matrix(10:13,2); det(A) # det= -2
B=rbind(A[2,],A[1,]) #row 2 first
B #view the interchanged rows
det(B) #this is +2
```

```
> A=matrix(10:13,2); det(A) # det= -2
[1] -2
> B=rbind(A[2,],A[1,]) #row 2 first
```

```
> B #view the interchanged rows
     [,1] [,2]
[1,]   11   13
[2,]   10   12
> det(B) #this is +2
[1] 2
```

4] If B is obtained from A by multiplying all the elements of only one row (or column), by a scalar constant k, then $det(B) = k\,det(A)$.

This property of determinants is different from an analogous property for matrices. If A is a matrix, then kA means multiply each element of A by the scalar k. However, in computing $k\,det(A)$ only one row or column is multiplied by k. Recall that this also follows from the expansions in Eq. (6.7).

Note that $det(kA) = k^n det(A)$ holds true by definition of scalar multiplication. Now choosing the scalar $k = -1$, we have the general result: $det(-A) = (-1)^n det(A)$.

5] If B is obtained from A by multiplying the elements of i-th row (or column) of A by a constant k, and then adding the result to the j-th row of A, then $det(B) = det(A)$. These operations are called elemetary row (or column) operations.

In our R example of a two dimensional matrix A let us choose $k = 7$. Thus, B is obtained by multiplying the elements of the first row by k and adding it to the second row (11, 13).

```
#R.snippet det(A) unchanged after elem. row operations.
A=matrix(10:13,2)
B=A  #initialize
w=7*A[1,] # 7 times first row or (70, 84)
B[2,]=B[2,]+w
B
det(B)  #should be -2
det(A) #has been -2 all along
```

Since the first row of A has (10, 12), 7 times it become (70, 84). Now we add this to the original second row.

```
> w=7*A[1,] # 7 times first row or (70, 84)
> B[2,]=B[2,]+w
> B
```

```
      [,1] [,2]
[1,]   10   12
[2,]   81   97
> det(B)   #should be -2
[1] -2
> det(A) #has been -2 all along
[1] -2
```

The example shows that the determinant has not changed.

6] Determinant of a diagonal matrix is simply the product of diagonal entries. For example, define a diagonal matrix with diagonal entries (3, 6, 8) and check that the determinant is 144. Output of following snippet is suppressed for brevity.

```
#R.snippet det. of diagonal is product of those terms
A=diag(c(3,6,8)); A
det(A) #is 144
3*6*8 #also 144
```

7] Determinant is a product of eigenvalues. $det(A) = \Pi_{i=1}^{n}\lambda_i(A)$, where the concept and definitions of eigenvalues are discussed in detail in a separate chapter Chap. 9.

In the following R snippet we use a random sample of numbers from 11 to 11+15 to define a 4×4 matrix. Its eigenvalues are computed by the function 'eigen' in R. This is a powerful and useful function in R which provides both eigenvalues and eigenvectors. To access only the eigenvalues, we have to define an object 'ei' as the output of the function 'eigen' and then use 'ei$values' to extract them. Note that we are using a useful facility in R to access various outputs created by R functions and then further manipulate those outputs. Here the manipulation is that we want to multiply out all the eigenvalues. Another R function called 'cumprod' for cumulative multiplication of all elements sequentially cumulatively. Hence the last element created by 'cumprod' will the product of all components.

In the following snippet $det(A)$ is -1338 and the last element of cumprod is also the same, as predicted by matrix theory. Also, in the following snippet the functions 'set.seed' and the function 'sample' in R are used. The 'set.seed' allows us to set the seed of the random number generator, so that each time anyone and everyone gets exactly the same set of randomly chosen numbers. Here the random number generator is used by the R function 'sample' and it creates a random sample out of the indicated range of numbers. If we do not set the seed R sets it automatically, based on the local computer system clock. This means that every run will have a different random sample with distinct answers. It will be pedagogically difficult to work with.

```
#R.snippet det(A) = product of its eigenvalues
set.seed(45); A=sample(11:(11+15)); A
A=matrix(A,4); A
det(A)
ei=eigen(A)
ei$values
cumprod(ei$values)
```

The output by R is given as

```
> A=matrix(A,4); A
     [,1] [,2] [,3] [,4]
[1,]   21   23   12   18
[2,]   15   24   11   26
[3,]   14   13   20   17
[4,]   25   22   19   16
> det(A)
[1] -1338
> ei=eigen(A)
> ei$values
[1] 74.6215875  9.6136276 -3.7346252  0.4994101
> cumprod(ei$values)
[1]    74.62159   717.38416 -2679.16098 -1338.00000
```

8] The determinant of an upper or lower triangular matrix is the product of its diagonal elements.

Definition: Triangular matrix:

An upper triangular matrix is one which contains nonzero elements only above and along the main diagonal. A lower triangular matrix is one which contains nonzero elements only below and along the main diagonal. Given a square matrix, the function 'triang' in the 'fBasics' package, Wuertz and core team (2010) in R extracts the lower triangular matrix, whereas the function 'Triang' with upper case 'T' extracts the upper triangular matrix. In the snippet we use a simple example.

```
#R.snippet det(triangular A) equals product of its diagonals
```

```
A=matrix(1:9,3) #create simple matrix 3 by 3
A #display the matrix
diag(A)#display diagonals
cumprod(diag(A)) #product of diagonals
triang(A)#display lower triangular matrix
det(triang(A)) #det of lower triangular matrix
Triang(A)#display upper triangular matrix
det(Triang(A))#det of upper triangular matrix
```

In the snippet the product of diagonals is computed by using the 'cumprod' function of R. The last term equals 45 and is the product of diagonal entries (1, 5, 9).

```
> A #display the matrix
     [,1] [,2] [,3]
[1,]   1    4    7
[2,]   2    5    8
[3,]   3    6    9
> diag(A)#display diagonals
[1] 1 5 9
> cumprod(diag(A)) #product of diagonals
[1]   1   5 45
> triang(A)#display lower triangular matrix
     [,1] [,2] [,3]
[1,]   1    0    0
[2,]   2    5    0
[3,]   3    6    9
> det(triang(A)) #det of lower triangular matrix
[1] 45
> Triang(A)#display upper triangular matrix
     [,1] [,2] [,3]
[1,]   1    4    7
[2,]   0    5    8
[3,]   0    0    9
> det(Triang(A))#det of upper triangular matrix
[1] 45
```

The output displays that a lower-triangular matrix obtained by the function triang (with lower case t), where all entries above the diagonal are made zero. Its determinant is seen to be $det(A) = 45$. Similarly the function Triang (with upper case T), where all entries below the diagonal are made zero. Its determinant is also seen to be $det(A) = 45$. This example illustrates the property that determinant of the upper triangular matrix is a product of its diagonal elements (1*5*9 =45).

6.3   Cramer's Rule and Ratios of Determinants

Gabriel Cramer's rule from 1750 (See http://en.wikipedia.org/wiki/ Cramer's_rule) is a clever method for solving a system of linear equations when the number of equations equals number of unknowns. It is valid whenever a unique solution does exist.

Consider a system of 2 equations

$$ax + by = e \qquad (6.9)$$
$$cx + dy = f$$

in 2 unknowns (x, y), where a, b, c, d, e, f denotes known numbers.

In matrix notation we write Eq. (6.9) as

$$\begin{bmatrix} a & b \\ c & d \end{bmatrix} \begin{bmatrix} x \\ y \end{bmatrix} = \begin{bmatrix} e \\ f \end{bmatrix}, \quad \text{that is,} \quad Az = r, \qquad (6.10)$$

where A is 2×2 matrix and where $z = (x, y)$, $r = (e, f)$ are 2×1 column vectors. Note that the determinant of A is $(ad - bc)$. This will be used in Cramer's rule denominators.

Cramer's rule involves replacing various columns of A by the r vector on the right hand side Eq. (6.10). Denote by A_1 the matrix obtained by replacing the first column of A by the 2×1 vector $r = (e, f)$. Similarly, obtain the A_2 matrix by replacing the *second* column of A by $r = (e, f)$.

Thus Cramer's rule suggests creating

$$A_1 = \begin{bmatrix} e & b \\ f & d \end{bmatrix} \text{ and } A_2 = \begin{bmatrix} a & e \\ c & f \end{bmatrix}. \qquad (6.11)$$

Once these matrices are created, Cramer's rule suggests writing the solution as a ratio of two determinants:

$$x = \frac{det(A_1)}{det(A)} = \frac{ed - fb}{ad - bc} \qquad (6.12)$$
$$y = \frac{det(A_2)}{det(A)} = \frac{af - ce}{ad - bc}$$

Of course, Cramer's Rule will fail when the denominator is zero, that is when A is singular as discussed in the next section. This method generalizes to $k > 2$ equations in a natural way. We simply have to create matrices $A_1, A_2, \ldots, A_k$, by replacing columns of the $k \times k$ matrix A by the $k \times 1$ column vector on the right hand side r of the system of k equations.

Exercise Use Cramer's Rule to solve the system of 3 equations in 3 unknowns $Az = r$ where $z = (x_1, x_2, x_3)$. Use seed 124 and sample nine numbers from 1 to 20 to create the A matrix on the left side and a sample of 3 numbers from 3 to 10 for the right hand side vector r.

```
# R program snippet 6.3.1 Cramer's Rule.
set.seed(124)
A=matrix(sample(1:20, 9),3,3); A
r=sample(3:10,3); r
A1=cbind(r,A[,2], A[,3]); A1
A2=cbind(A[,1],r,A[,3]); A2
A3=cbind(A[,1],A[,2],r); A3
x1=det(A1)/det(A)
x2=det(A2)/det(A)
x3=det(A3)/det(A)
rbind(x1,x2,x3) #the answer
solve(A) %*% r #alternate answer
```

The output of snippet 6.3.1 is given below. It begins with the A matrix and r vector and shows how A_1 to A_3 matrices are created and eventually how the ratios of determinants finds the solution for x_1, x_2, x_3.

```
> A=matrix(sample(1:20, 9),3,3); A
     [,1] [,2] [,3]
[1,]    2    7    9
[2,]    8    4   17
[3,]   10    5   12
> r=sample(3:10,3); r
[1] 5 8 9
> A1=cbind(r,A[,2], A[,3]); A1
     r
[1,] 5 7  9
[2,] 8 4 17
[3,] 9 5 12
> A2=cbind(A[,1],r,A[,3]); A2
       r
```

```
[1,]   2 5  9
[2,]   8 8 17
[3,]  10 9 12
> A3=cbind(A[,1],A[,2],r); A3
          r
[1,]   2 7 5
[2,]   8 4 8
[3,]  10 5 9
> x1=det(A1)/det(A)
> x2=det(A2)/det(A)
> x3=det(A3)/det(A)
> rbind(x1,x2,x3) #the answer
         [,1]
x1 0.5630631
x2 0.4144144
x3 0.1081081
> solve(A) %*% r
          [,1]
[1,]  0.5630631
[2,]  0.4144144
[3,]  0.1081081
```

The last few lines of the output of snippet 6.3.1 show that an alternate way of finding the solution of the system of 3 equations is from the matrix multiplication of A^{-1} with r. It is implemented by the R code 'solve(A)%*% r', to be discussed later in Sec. 8.5 after matrix inversion is explained.

6.4   Zero Determinant and Singularity

A matrix is singular when its determinant is zero. Hence the important issue is to determine when the $det(A) = 0$. There are four ways in which a matrix A can be singular with $det(A) = 0$.

1] $det(A) = 0$ when two rows of A are identical.

```
#R.snippet Zero determinant of singular matrix
set.seed(45); A=sample(11:(11+15)); A
A=matrix(A,4); B=A #initialize
B[1,]=A[2,]; B
det(B)
```

The snippet is designed to force two rows of B to be identical, namely the first two rows.

```
> B[1,]=A[2,]; B
     [,1] [,2] [,3] [,4]
[1,]   15   24   11   26
[2,]   15   24   11   26
[3,]   14   13   20   17
[4,]   25   22   19   16
> det(B)
[1] 0
```

The output shows that the determinant of a matrix with two identical rows is indeed zero.

2] $det(A) = 0$ when two columns of A are identical.

```
#R.snippet identical columns imply zero det.
set.seed(45); A=sample(11:(11+15)); A
A=matrix(A,4); B=A #initialize
B[,2]=A[,1]; B  #first two columns identical
det(B)
```

In the output by R, see that the first column of B equals the second column of B.

```
B
     [,1] [,2] [,3] [,4]
[1,]   21   21   12   18
[2,]   15   15   11   26
[3,]   14   14   20   17
[4,]   25   25   19   16
> det(B)
[1] -9.15934e-14  #this is as good as zero
```

Since first two columns of B are identical to each other the determinant of B must be zero, as is verified from the R output.

3] If any row or column of a matrix has all zeros, its determinant is zero.

```
#R.snippet All-zero row or column means zero determinant.
set.seed(45); A=sample(11:(11+15)); A
A=matrix(A,4); B=A #initialize
```

```
B[1,]= rep(0, ncol(B)); B
det(B)
```

Note that the *B* matrix as understood by R has the first row with all zeros.

```
> B
     [,1] [,2] [,3] [,4]
[1,]    0    0    0    0
[2,]   15   24   11   26
[3,]   14   13   20   17
[4,]   25   22   19   16
> det(B)
[1] 0
```

Thus we have an illustration for the result that 'all zero row' spells a zero determinant.

4] If one eigenvalue is zero, determinant is zero. Recall from property 7] in Sec. 6.2 above that the determinant is a (cumulative) product of all eigenvalues. When one eigenvalue is zero, the product is obviously zero.

This property can be verified in R by noting that one of the the eigenvalues for each of the singular matrices is zero (or extremely small).

```
#R.snippet above B with one zero eigenvalue means zero det(B)=0.
eib=eigen(B)
eib$values
```

The output by using the powerful R function 'eigen' gives both eigenvalues and eigenvectors. Since we are not interested in eigenvectors here, we define an object 'eib' inside R and then use the command 'eib$values' to see the eigenvalues. Note that 'eigen' orders them from the largest to the smallest (in absolute values) so that the last eigenvalue is zero for *B*.

```
[1] 56.383918  9.628390 -6.012308  0.000000
```

Singular matrices with zero determinant are important because the inverse (reciprocal) of a singular matrix does not exist. This property of singular matrices is an extension of a similar property for scalars, namely, non existence of division of ordinary numbers by a zero. Note that all elements of a singular matrix need not be zero.

6.4.1 *Nonsingularity*

Matrix A is nonsingular if $det(A) \neq 0$. Our illustrative four dimensional matrix A is nonsingular, since its determinant is nonzero.

```
#R.snippet nonsingular random matrix illustrated
set.seed(45); A=sample(11:(11+15)); A
A=matrix(A,4);
det(A) #-1338 is nonzero
```

We shall see that non-singularity is an important and often desirable property of a matrix, allowing us to invert the matrix.

Chapter 7

The Norm, Rank and Trace of a Matrix

We now consider three scalars or 1×1 entities called norm, rank and trace associated with matrices. These were mostly omitted in earlier discussion for lack of adequate machinery. A fourth entity called quadratic form is also 1×1 and will be discussed in Chapter 10

7.1 Norm of a Vector

A real vector is viewed as an arrow from the origin to a point in $\Re^n$. The word norm in the present context means size. The norm of vector x denoted by $\| x \|$ as a measure of size, is basically the Euclidean length of the vector. However, it is possible to think of other notions of size, provided they satisfy some fundamental properties of distances.

(a) Norm cannot be negative.
(b) Norm should be zero when the vector is a null vector.
(c) If we double all elements of the vector, the norm should double. That is, given a scalar w,

$$\| w\,x \| = | w | \| x \| . \tag{7.1}$$

(d) Similar to all distances, it should satisfy the triangular inequality that the sum of two sides must exceed that of one side. If y is a second vector then

$$\| x + y \| \leq \| x \| + \| y \| . \tag{7.2}$$

Note that the inequality becomes equality when the vectors are linearly dependent such that one is a nonnegative multiple of the other.

A norm of a vector can be defined from its inner product.

7.1.1   *Cauchy-Schwartz Inequality*

Let p and q be two vectors with same dimension and let $|<p \bullet q>|$ denote the absolute value of their Euclidean inner product defined from Eq. (3.2). Let $\| p \|$ denote the Euclidean norm of p and similarly for the norm of q defined in Eq. (3.3). Now we have the famous inequality:

$$|<p \bullet q>| \leq (\| p \| \| q \|) \tag{7.3}$$

A more general version of the inequality admitting complex numbers will be discussed later in Sec. 11.1. The following snippet attempts to check the inequality using R.

```
# R program snippet 7.1.1.1 is next.
D=1:5 #assumed to be column vector
E=11:15 #column vector
LHS=abs(t(D) %*% E) #left side
RHS=sqrt(sum(D^2))*sqrt(sum(E^2))
LHS #left side of C-S inequality
RHS #right side of Cauchy-Schwartz
```

Note that the left side is smaller than right side.

```
> LHS #left side of C-S inequality
      [,1]
[1,]   205
> RHS #right side of Cauchy-Schwartz
[1] 216.8525
```

Rao and Rao (1998, Sec. 16.2) discusses matrix trace versions of Cauchy-Schwartz inequality involving $B \in M_n$ and $A \in M_{n,m}$ such that vector space for A is a subset of the vector space for B, where $Z \in M_{n,m}$ is arbitrary and where B^+ denotes the Moore-Penrose generalized inverse of B. For example, $(tr(Z'A))^2 \leq (trZ'BZ)(trA'B^+A)$, where the inequality becomes equality if and only if BZ is proportional to A. There are a few more such inequalities.

7.2   Rank of a Matrix

An $T \times p$ matrix X is said to be of rank p if the dimension of the largest nonsingular square submatrix is p. Recall that nonsingular matrix means that its determinant is nonzero. In applied statistics including econometrics X is often a large data matrix of regressors with T observations arranged in p columns for p regressors. The idea of the 'rank of a matrix' is intimately related to linear independence mentioned in Sec. 3.2.1 as follows. Those tests for independence of vectors are applied to row vectors and/or column vectors comprising a matrix in this chapter.

We have

$$rank(A) = min[Row\ rank(A), Column\ rank(A)], \qquad (7.4)$$

where 'Row rank' is the largest number of linearly independent rows, and where 'Column rank' is the largest number of linearly independent columns.

Consider the following example:

$$A = \begin{bmatrix} a_{11}\ a_{12}\ a_{13} \\ a_{21}\ a_{22}\ a_{23} \end{bmatrix} = \begin{bmatrix} 15\ 30\ 53 \\ 25\ 50\ -2 \end{bmatrix} \qquad (7.5)$$

Note that $a_{12} = 2a_{11}$ and $a_{22} = 2a_{21}$. Hence the first two columns are linearly dependent. Hence, of the three columns, only two columns are linearly independent columns. The column rank of A is only 2, not 3.

Recall the definition that a set of vectors a_1, $a_2 \cdots, a_n$ is linearly dependent, if a set of scalars c_i exists which are not all zero and which satisfy the test: $\sum_{i=1}^{n} c_i a_i = 0$.

In the above example of a 2×3 matrix A we can define two column vectors: $a_1 = \begin{bmatrix} 15 \\ 25 \end{bmatrix}$ and $a_2 = \begin{bmatrix} 30 \\ 50 \end{bmatrix}$. Now we wish to use the test stated above to determine if a_1 and a_2 are linearly dependent. It is possible to choose the two constants as $c_1 = 2$ and $c_2 = -1$. Now it is easy to verify that the summation $(c_1 a_1 + c_2 a_2) = 0$, Since this is zero, the columns of A are linearly dependent.

In R, the rank is obtained readily by the function 'rk' available in the package called 'fBasics' as follows

```
# R snippet 7.2.1 using rk is next.
library(fBasics)  #assuming you have downloaded this package
A = c(15,25,-12,22,53,-2)
A=matrix(A,2,3)
rk(A)  # R output (omitted for brevity) states rank = 2
```

A square nonsingular matrix B satisfies:

$$rank(B) = row - rank(B) = column - rank(B). \qquad (7.6)$$

```
# R snippet 7.2.2 using rk function of fBasics package.
library(fBasics)  #assuming you have downloaded this package
set.seed(45); A=sample(11:(11+15)); # a square matrix
A=matrix(A,4);
rk(A)  # omitted R output says that the rank is 4
```

7.3 Properties of the Rank of a Matrix

Following properties of rank are useful.

1] If X is a $T \times p$ matrix, then a basic rank inequality states that:

$$rank(X) \leq min(T,p), \tag{7.7}$$

which says that the rank of a matrix is no greater than the smaller of the two dimensions of rows (T) and columns (p).

In Statistics and Econometrics texts one often encounters the expression that "the matrix X of regressors is assumed to be of **full** (column) rank." Since the number of observations in a regression problem should exceed the number of variables, $T > p$ should hold, as a practical matter, anyway. After all, one cannot have less observations T than the number of parameters p to estimate. Hence $min(T,p) = p$ in these applications will always be p, the number of columns. Therefore, $rank(X) \leq p$. If, in particular, $rank(X) = p$, the rank for the data in the regression problem will be the largest (best) it can be, and hence we say that X is of 'full (column) rank.' On the other hand, if $rank(X) < p$, then ordinary least squares estimator is not unique.

2] The rank of a transpose:

$$rank(A) = rank(A'). \tag{7.8}$$

In the following snippet, The R function 't(A)' means transpose of A, and 'rk(A)' means rank of A.

```
# R snippet 7.2.3 using rk is next.
library(fBasics)  #assuming you have downloaded this package
set.seed(45); A=sample(11:(11+15)); # a square matrix
A=matrix(A,4);
rk(A)  # omitted R output says that the rank is 4
rk(t(A)) #R output says that the rank is 4
```

3]

$$rank(A + B) \leq rank(A) + rank(B). \tag{7.9}$$

In the following snippet we have an example where we find that the rank on the left side is 4, which less than the sum of the ranks, which is 8.

```
# R snippet 7.2.4 Rank(A+B) less than Rank(A) +Rank(B)
library(fBasics)  #assuming you have downloaded this package
set.seed(45); A=sample(11:(11+15)); A  # a square matrix
A=matrix(A,4); B=sample(55:(55+15)); B=matrix(B,4)
rk(A); rk(t(A)) #both are 4
rk(A+B) #this is 4
rk(A)+rk(B) # this is 8
```

4] Sylvester's Inequality: If A is $m \times n$ and B is $n \times q$ rectangular matrices

$$rank(A) + rank(B) - n \le rank(AB) \le min[rank(A), rank(B)]. \quad (7.10)$$

Let us use the example of the snippet above to get a hands-on experience for Sylvester's inequality. Let us write the left hand side of the inequality as LHS and the first right side as RHS1 and the second right side as RHS2. Note that the R function 'ncol' counts the number of columns in a matrix. Thus 'ncol(A)' is n.

```
# R snippet 7.2.5 Sylvester inequality bounds for Rank(AB)
LHS=rk(A)+rk(B)-ncol(A); LHS  #this is 4
RHS1=rk( A%*%B); RHS1 #this is also 4
RHS2= min(rk(A), rk(B) ); RHS2 #this is also 4
```

5] If AB and BA are conformable and B is nonsingular, $rank(AB) =$ rank(BA) =rank(A).

In the following snippet we use A and B as the two 4 dimensional matrices (same as above) and first check whether the statement holds true. Next, we make the first row of B as all zeros, making B a singular matrix. We want to see what happens to the above relation. With a row of all zeros the rank of B is reduced by unity (no longer 4 but reduced to 3). Hence the ranks of the products with B, namely AB and BA are also reduced to only 3.

```
#R.snippet: Rank reduction upon multiplication by singular matrix.
rk(A %*% B); rk(B %*% A) #both are 4
B[1,]=rep(0, ncol(B)) # rep means repeat zeros
rk(A %*% B); rk(B %*% A) #both become 3
```

6] If A is a matrix of real numbers, and if A' denotes its transpose, then:
rank(A) =rank$(A'A)$=rank(AA').

```
# R snippet 7.2.6 Rank(A)=Rank(A Atranspose).
library(fBasics)  #assuming you have downloaded this package
set.seed(45); A=sample(11:(11+15)); A  # a square matrix
A=matrix(A,4);
rk(A)  # R output say that the rank is 4
rk(A %*% t(A) )  # is 4
rk(t(A) %*% A ) # is also 4
```

7] Assuming the matrix products AB, BC and ABC are well defined, we have the following inequality
rank(AB)+rank$(BC) \leq$ rank(B)+rank(ABC).

In the following snippet we will use the A and B as 4 dimensional matrices as above and use a new C matrix which is 4×3. The above inequality will then require that $7 \leq 7$.

```
# R snippet 7.2.7 rk(AB)+rk(BC) less than rk(B)+rk(ABC).
library(fBasics)  #assuming you have downloaded this package
set.seed(45); A=sample(11:(11+15)); A  # a square matrix
A=matrix(A,4); B=sample(55:(55+15)); B=matrix(B,4)
C=sample(3: (3+11)); C=matrix(C,4,3); C
rk(A%*%B) # this is 4
rk(B %*%C) #this is 3
rk(B)#this is 4
rk(A%*%B %*%C) #this is 3
```

8] The rank of a matrix equals the number of nonzero eigenvalues (this is a technical concept to be discussed later in a separate chapter with all relevant details) of a matrix. With the advent of reliable R computer programs for eigenvalue computation, it is easy to determine the rank by simply counting the number of nonzero eigenvalues.

The R function 'eigen' computes the eigenvalues and eigenvectors. In the following snippet 'ei=eigen(A%*%B)$val' creates the object named 'ei' which uses the dollar symbol to extracts only the eigenvalues of the matrix expression obtained by row-column multiplication. Now we want to compute the number of nonzero eigenvalues. The function 'length' computes the length of any vector. The symbol '!=' is a logical expression in R for the inequality $\neq$. The subsetting of the 'ei' object in R is interesting to learn. 'ei[ei!=0]' creates a vector of nonzero eigenvalues.

The function 'eigen' in R can be used only for square matrices. If we have a rectangular matrix such as BC above, then we have to use 'singular value decomposition' function called 'svd' in R. It gives 3 matrices whereby the rank is given by the number of nonzero singular values. The command 'sv=svd(B %*%C)$d' captures the singular values, because the R function 'svd' lists the singular values under the name 'd.' We count the number of nonzero ones by the subset function 'sv[sv!=0].'

```
# R snippet 7.2.8 rank equals number of nonzero eigenvalues.
ei=eigen(A%*%B)$val
length(ei[ei!=0]) # this is 4 as above.
sv=svd(B %*%C)$d
length(sv[sv!=0]) #this is also 3 as above
```

Note that the rank of AB is 4, but the rank of BC is 3.

7.4 Trace of a Matrix

Trace of a matrix is simply the summation of its diagonal elements. Trace also equals the sum of eigenvalues.

$$Tr(A) = \sum_{i=1}^{n} a_{ii} = \sum_{i=1}^{n} \lambda_i(A), \tag{7.11}$$

where a_{ii} denotes the i-th diagonal value of A, and where $\lambda_i(A)$ denotes i-th eigenvalue of A (to be defined more precisely in a Chap. 9).

Matrix theory proves two further equalities regarding the trace of the sum and product of two conformable matrices:

$$Tr(A + B) = Tr(A) + Tr(B), \text{ and } Tr(AB) = Tr(BA). \tag{7.12}$$

In statistics and economics one often uses the latter result to simplify matrix expressions. For example, in regression models the so called hat matrix is defined as $X(X'X)^{-1}X'$. Equation (9.25) explains placing a hat on y through this matrix. Now let us choose $A = X$ and $B = (X'X)^{-1}X'$ and apply Eq. (7.12). $Tr(AB) = Tr(BA)$ now means that we can relocate the X at the end and conclude: $Tr(X(X'X)^{-1}X') = Tr((X'X)^{-1}X'X) = Tr(I_p) = p$. Since this trace equals the trace of an identity matrix, (sum of ones) the trace ultimately equals the column dimension p of X. This happens to be a major simplification. The point is that purely algebraic results can be useful in practice and hence are worth learning.

The following snippet writes a new *ad hoc* function in R to compute the trace of a matrix from summing the diagonals. It prints the dimensions of the matrix to the screen to make sure that the input A to the function 'mytr' for my trace is a proper matrix with two dimensions. Only then it computes the sum of diagonals and 'returns' that sum as the output of the new R function.

```
# R snippet 7.4.1 tr(A+B)=tr(A)+tr(B), tr(AB)=tr(BA).
set.seed(45); A=sample(11:(11+15)); # A is 4 by 4 square matrix
A=matrix(A,4); B=sample(55:(55+15)); B=matrix(B,4)
mytr=function(A){#new R function called mytr=my trace
print(c("dimension", dim(A))) #print dimensions,
#Ensure that A is a matrix
return(sum(diag(A))) } #function mytr ends here
ApB=A+B; ApB#view A+B
mytr(ApB) #use mytr to compute left side
mytr(A)+mytr(B)#right side
mytr(A%*%B) #second left side
mytr(B%*%A) #second right side
```

As one can see the left side for the first equation in Eq. (7.12) equals 347 and right side also equals 347 in the R output below. We also use the new function 'mytr' for computing the trace and check the two sides the second equation in Eq. (7.12). They are called in the snippet 7.4.1 as 'second left side' and 'second right side'. In its output below, they are both found to equal 18597.

```
> mytr=function(A){#new R function called mytr=my trace
+ print(c("dimension", dim(A))) #print dimensions,
#Ensure that A is a matrix
+ return(sum(diag(A))) } #function mytr ends here
> ApB=A+B; ApB#view A+B
      [,1] [,2] [,3] [,4]
[1,]   91   79   74   87
[2,]   72   91   66   87
[3,]   74   81   83   76
[4,]   83   86   84   82
> mytr(ApB) #use mytr to compute left side
[1] "dimension" "4"          "4"
```

```
[1] 347
> mytr(A)+mytr(B)#right side
[1] "dimension" "4"          "4"
[1] "dimension" "4"          "4"
[1] 347
> mytr(A%*%B) #second left side
[1] "dimension" "4"          "4"
[1] 18597
> mytr(B%*%A) #second right side
[1] "dimension" "4"          "4"
[1] 18597
```

7.5   Norm of a Matrix

In Sec. 7.1 we considered the norm of a vector as its size measured by its Euclidean length. Similar to Sec. 7.1 we list some definitional properties of the norm of a square $n \times n$ matrix denoted by $\|A\|$, as a nonnegative function of A satisfying following properties (given that B is a square $n \times n$ matrix also):

(1) $\|A\| \geq 0$ and $\|A\| = 0$, iff $A = 0$
(2) $\|cA\| = |c|\|A\|$, when c is a scalar
(3) $\|A + B\| \leq \|A\| + \|B\|$
(4) $\|AB\| \leq \|A\|\|B\|$

A general norm of a matrix is sometimes given in terms of the conjugate transpose A^* by

$$\|A\| = tr(A^*A) = tr(AA^*) \tag{7.13}$$

which admits complex numbers as elements of A. If the matrix has only real values, then Eq. (7.13) simplifies as $\|A\| = tr(A'A) = tr(AA')$.

The Euclidean norm of a square matrix A is simply the vector norm after the matrix is vectorized. It can be directly defined with subscript E as:

$$\|A\|_E = \left[\sum_i \sum_j (|a_{ij}|)^2 \right]^{1/2} . \tag{7.14}$$

The one norm is the maximum of sums of absolute values of elements in the column of A defined with subscript 1 as:

$$\|A\|_1 = max_j \left[\sum_{i=1}^{n} |a_{ij}| \right]. \tag{7.15}$$

The two norm (or spectral norm) is the square root of the maximum eigenvalue of $A'A$ defined with subscript 2 as:

$$\|A\|_2 = [\lambda_{max}(A'A)]^{1/2}. \tag{7.16}$$

The infinity norm is the maximum of sums of absolute values of row elements defined with subscript ∞ as:

$$\|A\|_\infty = max_i \left[\sum_{j=1}^{n} |a_{ij}| \right]. \tag{7.17}$$

The M norm is the maximum modulus of all elements of A defined with subscript M as:

$$\|A\|_M = max_j [|a_{ij}|]. \tag{7.18}$$

Some discussion is available at `http://en.wikipedia.org/wiki/Matrix_norm`, where matrix norms are defined in terms of singular values discussed later in Chapter 12.

The 'norm' function in the 'base' package of R computes all these norms above. The same norms are also computed by using the R package 'Matrix' Bates and Maechler (2010). We illustrate computation of various norms with a simple 3×3 matrix.

```
# R program snippet 7.5.1 matrix norms.
set.seed(992)
x=matrix(sample(1:9), 3,3)
x #prints the 3 by 3 matrix
norm(x, "1")#one norm max absolute col sum
norm(x, "I")#infinity norm max abs row sum
norm(x, "M")#M norm max modulus
norm(x, "F")#Euclidean or Frobenius norm
y=matrix(x,9,1) # vectorize x as 9 by 1
sqrt(sum(y^2))# should equal Euclidean norm
ei=eigen(t(x) %*% x)
sqrt(max(ei$val))#2 norm=sqrt(max eigenvalues)
```

With the simple matrix used int the snippet 7.5.1 the reader can visually verify in its output below that the norm definitions are correctly followed by the R software. For the Euclidean norm, we provide an extra line

`` `sqrt(sum(y^2))'``

in the snippet 7.5.1 for verification. The '2 norm' of Eq. (7.16) is not directly given by R, but it can be easily computed from the R function 'eigen' used before. Note that 'ei$val' reports all eigenvalues, the R function 'max' computes their maximum, and the R function 'sqrt' computes the square root. This is shown in the snippet 7.5.1.

```
> x #prints the 3 by 3 matrix
     [,1] [,2] [,3]
[1,]    6    3    4
[2,]    7    8    5
[3,]    9    2    1
> norm(x, "1")#one norm max absolute col sum
[1] 22
> norm(x, "I")#infinity norm max abs row sum
[1] 20
> norm(x, "M")#M norm max modulus
[1] 9
> norm(x, "F")#Euclidean or Frobenius norm
[1] 16.88194
> y=matrix(x,9,1) # vectorize x as 9 by 1
> sqrt(sum(y^2))# should equal Euclidean norm
[1] 16.88194
> ei=eigen(t(x) %*% x)
> sqrt(max(ei$val))#2 norm=sqrt(max eigenvalues)
[1] 16.06877
```

The above example shows that R software has made computation of matrix norms extremely easy.

The **spectral radius** of an $n \times n$ matrix is defined as its largest absolute eigenvalue:

$$\rho_s(A) = max(|\lambda_i|) \quad (i = 1, 2, \ldots n). \tag{7.19}$$

It is tempting to propose ρ_s as a measure of the size of matrix, but it can be zero without A being zero, such as when A is 2×2 matrix with all zeros except for unity in the top right corner. Another problem is that it fails a fundamental property of distances in Eq. (7.2), the triangle inequality: $\rho_s(A + B) > \rho_s(A) + \rho_s(B)$, for the above 2×2 matrix A and choosing B as its transpose. However the spectral radius of a square matrix remains a useful concept, because it provides a *conservative* lower limit to its size measured by *any* norm:

$$\rho_s(A) \leq \| A \| . \tag{7.20}$$

Chapter 8

Matrix Inverse and Solution of Linear Equations

One of the motivations for learning matrix theory is solving a system of equations similar to Eq. (3.8). This is a very old problem in mathematics and most textbooks on matrix algebra discuss several older elimination methods, which do not necessarily involve computation of the matrix inverse. Some of the older methods for solving equations are unnecessarily cumbersome and need not be discussed in a modern computing environment. This chapter limits the discussion of such methods to the bare minimum.

8.1 Adjoint of a Matrix

Let A_{ij} denote an $(n–1)\times(n-1)$ sub-matrix (called a minor) obtained by deleting i-th row and j-th column of an $n \times n$ matrix $A = (a_{ij})$. The cofactor of A is defined as:

$$C = \{c_{ij}\}, \text{ where } c_{ij} = (-1)^{i+j} det(A_{ij}). \tag{8.1}$$

Cofactors are also mentioned in Sec. 6.1.

The adjoint of A is defined as:

$$\text{Adj}(A) = C' = \{c_{ji}\}, \tag{8.2}$$

which involves interchanging the rows and columns of the matrix of cofactors indicated by the subscript ji instead of the usual ij. The operation of interchange is called the transpose of the matrix and denoted by a prime symbol. This adjoint should not be confused with the complex adjoint in Sec. 4.2.

8.2   Matrix Inverse and Properties

The matrix inverse is defined for square matrices only and denoted by a superscript -1. If $A = \{a_{ij}\}$ with i,j= $1, 2, \cdots, n$ then its inverse is defined as:

$$A^{-1} = \{a^{ij}\} = \frac{Adj(A)}{det(A)} = (-1)^{i+j} \frac{det(A_{ji})}{det(A)}, \qquad (8.3)$$

where the element at the i,j location of A^{-1} is denoted by the superscript a^{ij}. Note also that the subscripts have a special meaning. The A_{ji} denotes the sub-matrix of A obtained by eliminating j-th row and i-th column. Since the denominator of the above formula for the inverse of a matrix A contains its determinant $det(A)$, the inverse of a matrix is not defined unless the matrix is nonsingular, i.e., has nonzero determinant. Otherwise, one encounters the impossible problem of dividing by a zero.

The above method of finding the inverse is not recommended on a digital computer due to rounding and truncation errors explained in McCullough and Vinod (1999). These errors are serious for large or ill-conditioned matrices, not for small illustrative examples used in textbooks. However, we do not know in advance when they might be serious and best avoided by using a numerically reliable algorithms in R. In the following snippet we create a random A matrix and compute its inverse in R by using the reliable 'solve' function in R. We immediately verify that it gave a correct inverse by checking whether $A^{-1}A = I = AA^{-1}$, where I is the identity matrix.

```
#R.snippet
set.seed(987); a=sample(5:20)
A=matrix(a[1:4],2,2); A #define 2 by 2 matrix
Ainv=solve(A); Ainv #view inverse matrix
round(Ainv%*%A,13) #should be identity matrix
#conclude that Ainv is indeed inverse of A
detA=det(A); detA #view determinant
adjo=detA*Ainv #get adjoint after inverse
adjo #view adjoint

> A=matrix(a[1:4],2,2); A #define 2 by 2 matrix
      [,1] [,2]
[1,]   12   13
[2,]   19   20
```

```
> Ainv=solve(A); Ainv #view inverse matrix
          [,1]      [,2]
[1,] -2.857143  1.857143
[2,]  2.714286 -1.714286
> round(Ainv%*%A,13) #should be identity matrix
     [,1] [,2]
[1,]   1    0
[2,]   0    1
> #conclude that Ainv is indeed inverse of A
> detA=det(A); detA #view determinant
[1] -7
> adjo=detA*Ainv #get adjoint after inverse
> adjo #view adjoint
      [,1] [,2]
[1,]   20  -13
[2,]  -19   12
```

Some Properties of Matrix Inverse:

1] Inverse of inverse gives back the original matrix $A = (A^{-1})^{-1}$.

```
#R.snippet using previously defined matrix A
solve(Ainv)
```

```
> solve(Ainv)
     [,1] [,2]
[1,]   12   13
[2,]   19   20
```

2] Inverse of transpose equals the transpose of the inverse: $(A')^{-1} = (A^{-1})'$.

```
#R.snippet using previously defined matrix A
AinvTr=solve(t(A));AinvTr #inv of transpose
TrAinv=t(Ainv);TrAinv#transpose of inverse
```

```
> AinvTr=solve(t(A));AinvTr #inv of transpose
          [,1]      [,2]
[1,] -2.857143  2.714286
[2,]  1.857143 -1.714286
> TrAinv=t(Ainv);TrAinv#transpose of inverse
          [,1]      [,2]
```

```
[1,] -2.857143  2.714286
[2,]  1.857143 -1.714286
```

3] Inverse of a product of matrices follows the following rule:

$$(AB)^{-1} = B^{-1}A^{-1}, \quad (ABC)^{-1} = C^{-1}B^{-1}A^{-1}, \qquad (8.4)$$

where the order is reversed, which is similar to Eq. (4.7) for the transpose of a product of two or more matrices. However, do not apply a similar rule to the inverse of a sum. In general, $(A+B)^{-1} \neq A^{-1}+B^{-1}$. In fact even if A^{-1} and B^{-1} exist their sum may not even have an inverse. For example, if $B = -A$ the sum $A+B$ is the null matrix, which does not have an inverse, since one cannot divide by a zero.

```
#R.snippet using previously defined matrix A
B=matrix(a[5:8],2,2);B#view B
AB=A%*%B; AB #view AB
Binv=solve(B); Binv#view B inverse
ABinv=solve(AB);ABinv#view AB inverse Left side
BinvAinv=Binv%*%Ainv;BinvAinv#view Right side
```

If instead of the correct $B^{-1}A^{-1}$, one uses $A^{-1}B^{-1}$, without reversing the order, the left side does not equal the right side, as it does below.

```
> AB=A%*%B; AB #view AB
       [,1] [,2]
[1,]   259   212
[2,]   406   331
> Binv=solve(B); Binv#view B inverse
            [,1]        [,2]
[1,]   0.1632653 -0.1836735
[2,]  -0.1428571  0.2857143
> ABinv=solve(AB);ABinv#view AB inverse Left side
            [,1]        [,2]
[1,]  -0.9650146  0.6180758
[2,]   1.1836735 -0.7551020
> BinvAinv=Binv%*%Ainv;BinvAinv#view Right side
            [,1]        [,2]
[1,]  -0.9650146  0.6180758
[2,]   1.1836735 -0.7551020
```

4] The determinant of a matrix inverse is the reciprocal of the determinant of the original matrix: $\det(A^{-1}) = 1/\det(A)$

```
#R.snippet using previously defined matrix A
det(Ainv) #left side
1/det(A) # right side
```

See that left side equals the right side as claimed by the property of matrices from theory.

```
> det(Ainv) #left side
[1] -0.1428571
> 1/det(A) # right side
[1] -0.1428571
```

5] If a diagonal matrix A has $a_1, a_2, \ldots, a_n$ along the diagonal, A^{-1} has $1/a_1, 1/a_2, \ldots, 1/a_n$ along its diagonal.

```
#R.snippet using previously defined matrix A
dA=diag(A); dA#extract diagonals
dAd=diag(dA); dAd#view matrix of diagonals
solve(dAd)#left side
recip=1/dA; recip;#reciprocals of diagonals
diag(recip)#right side=matrix of reciprocals
```

```
> dA=diag(A); dA#extract diagonals
[1] 12 20
> dAd=diag(dA); dAd#view matrix of diagonals
      [,1] [,2]
[1,]   12    0
[2,]    0   20
> solve(dAd)#left side
           [,1] [,2]
[1,] 0.08333333 0.00
[2,] 0.00000000 0.05
> recip=1/dA; recip;#reciprocals of diagonals
[1] 0.08333333 0.05000000
> diag(recip)#right side=matrix of reciprocals
           [,1] [,2]
[1,] 0.08333333 0.00
[2,] 0.00000000 0.05
```

8.3 Matrix Inverse by Recursion

If we want to compute the inverse of $(A - zI)^{-1}$ recursively let us write it as

$$(A - zI)^{-1} = -(zI - A)^{-1} = \frac{1}{\Delta}[Iz^{n-1} + B_1 z^{n-1} + \ldots + B_{n-1}], \quad (8.5)$$

where

$$\Delta = det(zI - A) = z^n + a_{n-1}z^{n-1} + \cdots + a_0, \quad (8.6)$$

is a polynomial in complex number z of the so-called z-transform. This is often interpreted as $z = L^{-1}$, where L is the lag operator in time series $(Lx_t = x_{t-1})$. Multiplying both sides by Δ we get polynomials in z on both sides. Equating the like powers of z we obtain the recursion:

$B_1 = A + a_{n-1} I$
$B_2 = AB_1 + a_{n-2} I$
$B_k = AB_k + a_{n-k}I$ for $k = 2, \ldots, n - 1$,

and finally, since B_n is absent above, we have
$B_n = 0 = AB_{n-1} + a_0 I.$

8.4 Matrix Inversion When Two Terms Are Involved

Let G be an $n \times n$ matrix defined by

$$G = [A + BDB']^{-1}, \quad (8.7)$$

where A and D are nonsingular matrices of order n and m, respectively, and B is $n \times m$. Then

$$G = A^{-1} - A^{-1}B[D^{-1} + B'A^{-1}B]^{-1}B'A^{-1}. \quad (8.8)$$

The result is verified directly by showing that $GG^{-1} = I$. Let $E = D^{-1} + B'A^{-1}B$. Then

$$G^{-1}G = I - BE^{-1}B'A^{-1} \quad (8.9)$$
$$+BDB'A^{-1} - BDB'A^{-1}BE^{-1}B'A^{-1}$$
$$= I + [-BE^{-1} + BD - BDB'A^{-1}BE^{-1}]B'A^{-1}$$
$$= I + BD[-D^{-1}E^{-1} + I - B'A^{-1}BE^{-1}]B'A^{-1}$$
$$= I + BD[I - EE^{-1}]B'A^{-1} = I,$$

where $(I - I) = 0$ eliminates the second term.

An important special case arises when $B = b$ is an $n \times 1$ vector, and $D = 1$. Then the inverse of two terms has a convenient form:

$$(A + bb')^{-1} = A^{-1} - \frac{A^{-1}bb'A^{-1}}{1 + b'A^{-1}b.} \qquad (8.10)$$

8.5   Solution of a Set of Linear Equations $Ax = b$

Section 6.3 explains Cramer's Rule for solving a set of k linear equations $Az = r$ in k unknowns when the solution is unique and the matrix A is nonsingular. In this section, let A be a 2×2 matrix defined as:

$$A = \begin{bmatrix} 5 & 3 \\ 4 & 2 \end{bmatrix} \qquad (8.11)$$

Now, let $x = (x_1, x_2)'$, be the column vector of unknown quantities to be computed (solved for) and let b= $(19, 14)'$ be the right hand known column vector of constants.

Thus $Ax = b$ represent the following system of two non-homogeneous equations in two unknowns x_1 and x_2.

$$5x_1 + 3x_2 = 19$$
$$4x_1 + 2x_2 = 14 \qquad (8.12)$$

Note that the right hand vector b is not all zeros. If, in particular, $b = 0$, the system is known as a homogeneous system. Note that the solution of a homogeneous system is indeterminate or non-unique, if we refuse to admit the trivial solution $x = 0$ as one of the solutions. This will be relevant in Chap. 9.

Now solving the system of two equations amounts to re-writing them with x on the left hand side. Pre-multiplying both sides of $Ax = y$ by A^{-1} we are left with only the 2×1 vector x on the left hand side, since $A^{-1}A = I$, the identity matrix (and $I\,x = x$ always holds). For our simple example, it is easy to verify that $x = A^{-1}b = (2,3)'$. That is, $x_1 = 2$ and $x_2 = 3$ are the solutions for the two unknowns.

Here is how we proceed in R, without bothering with cofactors or determinants. The determinant is only indirectly involved, because the inverse will not exist if the matrix is singular and the matrix is singular when the determinant is zero.

```
#R.snippet
A=matrix(c(5,4,3,2),2,2)
A #visualize the A matrix
Ainv=solve(A); Ainv# visualize inverse matrix
x=Ainv %*%  c(19,14) #post multiply by b
x # solution
```

Matrix inversion in R is obtained by a simple command 'solve.' Post-multiplication of the matrix inverse 'Ainv' by the column vector $b = (19, 14)$ is obtained by the R command 'Ainv %*% c(19,14)' in the snippet. Note that 'c(19,14)' implicitly defines a column vector in R.

```
> A #visualize the A matrix
      [,1] [,2]
[1,]    5    3
[2,]    4    2
> Ainv=solve(A); Ainv# visualize inverse matrix
      [,1] [,2]
[1,]   -1  1.5
[2,]    2 -2.5
> x=Ainv %*%  c(19,14) #post multiply by b
> x # solution
      [,1]
[1,]    2
[2,]    3
```

The solution of a system of equations is a very common problem arising in various garbs throughout the scientific literature, and matrix algebra offers a powerful tool for numerically obtaining the solutions.

For example, consider the least squares regression model mentioned in the Preface as Eq. (0.1): $y = X\beta + \epsilon$ in matrix notation, where there are p regressors and T observations. We shall see later in Sec. 8.9 that the first order conditions for minimization of error sum of squares ($\epsilon'\epsilon$) leads to the so-called 'normal equations' $(X'X)\beta = X'y$, which represent a system of p equations in p unknowns in the form of p regression coefficients in vector β. Of course, this system of equations does not look like $Ax = b$, until we realize the change of notation: $A = (X'X), x = \beta, b = (X'y)$.

8.6   Matrices in Solution of Difference Equations

Given a time series y_t, $(t = 1, 2, \ldots, T)$ when a relation involves some lagged values $y_{t-k} = L^k y_t$ having a power k of the lag operator L, we have a difference equation. For example, the first order stochastic difference equation is defined as:

$$y_t = \rho y_{t-1} + a_t, \quad a_t \sim N(0, 1) \tag{8.13}$$

where the relation has 'white noise' random input (i.e., from a Normal variable with mean 0 and variance unity). Its solution is obtained by inverting a polynomial in the lag operator, or writing Eq. (8.13) as

$$y_t = (1 - \rho L)^{-1} a_t = \sum_{j=0}^{\infty} G_j a_{t-j} \tag{8.14}$$

where we have an infinite series in the lag operator with terms $1, \rho L, (\rho L)^2, \ldots$. If we define $z = L^{-1}$, The coefficients $G_j = \rho^j$ contain ρ which is a root of the 'characteristic' polynomial $(1 - \rho L) = 0 = (z - \rho)$ in z.

Matrix algebra allows us to generalize this first order autoregressive AR(1) model to a 'vector' autoregressive, VAR(p), model having k variables and p lags as explained in Vinod (2008a, ch. 5). Note that VAR(p) specification stacks the k variables into one $k \times 1$ vector y_t and replaces the single coefficient ρ with $k \times k$ matrices $A_1, A_2, \ldots, A_p$. Thus, matrix algebra allows compact notation and helps in finding solutions.

Since the solution of AR(1) model is intuitive, it helps to rewrite the VAR(p) model as VAR(1) model by stacking the dependent variable y_t into $kp \times 1$ vector Y_t with k elements for each of the p lags: $Y_t = [y_t, y_{t-1}, \ldots, y_{t-p+1}]'$. Also, we stack adequate number of zero vectors below the error a_t to create a conformable $kp \times 1$ vector of errors a_t. Now, VAR(p) is equivalent to the following VAR(1):

$$Y_t = AY_{t-1} + a_t, \tag{8.15}$$

with a large $kp \times kp$ matrix A having the first row containing all the coefficient matrices A_1 to A_p associated with the p lags. For example, if we start with VAR(p) with $p = 3$, we can rewrite it as the following VAR(1) of Eq. (8.15):

$$\begin{pmatrix} y_t \\ y_{t-1} \\ y_{t-2} \end{pmatrix} = \begin{bmatrix} A_1 & A_2 & A_3 \\ I & 0 & 0 \\ 0 & I & 0 \end{bmatrix} \begin{pmatrix} y_{t-1} \\ y_{t-2} \\ y_{t-3} \end{pmatrix} + \begin{pmatrix} a_t \\ 0 \\ 0 \end{pmatrix}. \tag{8.16}$$

In general, A has a large $k(p-1) \times k(p-1)$ identity matrix along the $p-1$ columns of A and the last column has all zeros.

In the absence of white noise input a_t in Eq. (8.13) if we start with y_0 the solution is $y_t = \rho^t y_0$. Similarly, if the white noise input is absent, the solution of the general model in Eq. (8.15) is $Y_t = A^t Y_0$. One simply needs the powers of the A matrix. Eigenvalues of A discussed in Chapter 9 are useful in finding solutions for certain models of this type.

8.7   Matrix Inverse in Input-output Analysis

Wassily Leontief received the Nobel prize in Economics for a powerful tool called input-output analysis, which involves a matrix inverse, Leontief (1986). See also `http://en.wikipedia.org/wiki/Input-output_model` for some details regarding Leontief input-output analysis. The US benchmark tables are described in Stewart *et al.* (2007). An excellent reference including mathematical derivations is Horowitz and Planting (2006) (hereafter "HP06") showing exactly how it is used by US economic policy makers at the Bureau of Economic Analysis (BEA). Data for US is available at `http://www.bea.gov/industry/index.htm#benchmark_io`. Input output data for OECD and developing countries are also available.

Since HP06 has established a standard followed by many countries let us use their terminology. Their so-called **Use Table** shows the uses of commodities by intermediate and final users. The rows in the use table present the commodities or products, and the columns display the industries and final users that utilize them. The row sum is the commodity output, while columns show the products consumed by each industry. The "value added" has three components. (i) compensation of employees, (ii) taxes on production, and imports less subsidies, and (iii) gross operating surplus. Value added is the difference between an industry's output and the cost of its intermediate inputs. Note that total value added for all industries equals the gross domestic product (GDP). The column sum of 'use table' equals that industry's output. The data collection and reconciliation is an important task which is greatly helped by matrix algebra including input-output analysis.

Their so-called **Make Table** matrix shows the value in producers' prices of each commodity produced by each industry. The row entries equal the dollar value of commodities produced by the industry. The column entries equal the dollar value of production by each industry.

The economy is divided into n sectors or industries with row and column headings matching the name of the sector. Let x_{ij} represent the amount of the product of sector i the row industry needed as input to sector j the column industry. An entry on the i-th row and j-th column in the input-output coefficient matrix A is represented as $\{a_{ij}\} = x_{ij}/x_j$, where x_j represents total gross production of j-th industry. Let us use sample 'Use-Table' from Table 12.5 of HP06 to illustrate the tabulation of data.

\# R program snippet **8.7.1** is next.

```
A=c(50,120,120,40,330) #data Row for industry A
B=c(180, 30, 60,130, 400)#data Row for industry B
C=c(50, 150, 50, 20, 270)#data Row for industry C
scrap=c(1, 3, 1, 0, NA)
VaAdd=c(47, 109, 34, 190, NA)#value added
usetab=rbind(A,B,C,scrap,VaAdd)#create matrix
colsu=apply(usetab,2,sum)#col sum of matrix
usetable=rbind(usetab, colsu)#insert extra row
#provide column names FinDd=Final demand
colnames(usetable)=c("A", "B", "C", "FinDd", "Tot")
print("Sample Use Table 12.5 from HP06")
usetable #Table 12.5 in HP06
X=usetable[1:3,5]
xij=usetab[1:3,1:3]
A=xij %*% diag(1/X) #input-output coefficients
A
Finaldemand=X-A%*%X
cbind(Finaldemand, usetab[1:3,4])
```

Let X denote a column vector containing the row sums. In our illustration it is in the fifth column of the Use Table. The inter-industry transactions matrix x_{ij} is the top 3×3 submatrix of the Use Table. The command 'A=xij %*% diag(1/X)' in the snippet 8.7.1 creates the matrix of input-output coefficients representing input per unit of output in a 'cooking recipe' type linear production function. If a cook wants two servings and the recipe has measurements per serving, the ingredients per serving are simply doubled to get two servings. Input-output analysis assumes that the economy follows a similar linearity. My own Harvard dissertation attempted to relax this assumption by incorporating nonlinearities.

The output of the snippet 8.7.1 below reports the A matrix. For brevity, we omit discussion of the fourth row named 'scrap' used by HP06 in their 'Use Table' below. This entry permits some practical data adjustments and reconciliations. The matrix A cannot be any arbitrary matrix of real numbers, economists need all its element to be nonnegative.

The last line of the snippet 8.7.1 verifies that Final Demand obtained from the matrix expression 'X-A%*%X' equals the Final Demand abbreviated as 'FinDd' in the fourth column of the illustrative Use Table below.

```
> print("Sample Use Table 12.5 from HP06")
[1] "Sample Use Table 12.5 from HP06"
> usetable #Table 12.5 in HP06
          A   B   C FinDd Tot
A        50 120 120    40 330
B       180  30  60   130 400
C        50 150  50    20 270
scrap     1   3   1     0  NA
VaAdd    47 109  34   190  NA
colsu   328 412 265   380  NA
> X=usetable[1:3,5]
> xij=usetab[1:3,1:3]
> A=xij %*% diag(1/X) #input-output coefficients
> A
          [,1]   [,2]      [,3]
A 0.1515152 0.300 0.4444444
B 0.5454545 0.075 0.2222222
C 0.1515152 0.375 0.1851852
> Finaldemand=X-A%*%X
> cbind(Finaldemand, usetab[1:3,4])
  [,1] [,2]
A   40   40
B  130  130
C   20   20
```

Note that by definition,

$$X_t = AX_t + \tilde{F}_t, \qquad (8.17)$$

where Final Demand vector is denoted as $\tilde{F}$, and where we have inserted a time subscript t. The Eq. (8.17) can be made a difference equation by writing it as:

$$X_t = AX_{t-1} + \tilde{F}_t, \qquad (8.18)$$

where last year's output is linked to this year's output, making the input-output system dynamic. Now we can think of an equilibrium (steady state) output vector X_* which does not change over time, that is, we have:

$$X_* = AX_* + \tilde{F}, \tag{8.19}$$

where we want to know if $X_t \to X_*$ as $t \to \infty$, that is if the economy reaches an equilibrium.

Economic policy makers wish the specify the Final Demand vector and evaluate its effects on various sectors of the economy. For example, they may specify specified growth in Final Demand at levels (10%, 15%, 12%) of original values. The resulting Final Demand is denoted by 'F' and called 'enhanced Final Demand' in the sequel and in the following snippet 8.7.2. The policy maker is studying possible bottlenecks, that is, interested in deriving the required industry outputs in the revised X vector denoted as X_{new}, such that it can fulfill that enhanced Final Demand F.

This and a variety of important policy questions are formulated as derivation of the implied 'Requirements Table' induced by specified F. The idea is to study whether entries in the requirements table are feasible within the specified policy time horizon. For example, if the X_{new} requires that steel output be increased by 20%, one can check if such growth is feasible. Note that the answer is readily found by using matrix algebra inverse of the $(I - A)$ matrix as:

$$X_{new} = (I - A)^{-1}F = (I + A + A^2 + \ldots)F \ldots, \tag{8.20}$$

where we must assume that the inverse exists, that is the matrix is nonsingular, as discussed in Section 6.4.1.

Note that the last equation involving infinite series seen in Eq. (8.20) includes the matrix version of the usual scalar expansion $(1-a)^{-1} = 1+a+a^2+\ldots$ into infinite series. It is well known that the scalar series converges if $|a| < 1$ holds. In light of the question posed after Eq. (8.18) we can say that economy reaches an equilibrium (steady state) if *any* matrix norm satisfies condition $\|A\| < 1$. Otherwise the matrix infinite series in A^k is divergent.

The convergence condition in terms of eigenvalues $\lambda_i(A)$ is that the modulus of each eigenvalue is less than unity. Consider the A in snippet 8.7.1

```
# R program snippet 8.7.2 is next.
#earlier snippet must be in memory
ei=eigen(A)
abs(ei$val)#print absolute eigenvalues of A
```

[1] 0.8149689 0.2601009 0.2601009

The optput of snippet 8.7.2 shows that absolute values of all eigenvalues of the A of our example are indeed less than unity: $|\lambda_i(A)| < 1$.

8.7.1 *Non-negativity in Matrix Algebra and Economics*

Note that in pure matrix algebra, the solution X_{new} may well have some negative numbers. In Economic applications negative outputs do not make economic sense. Hence we need additional condition for the existence of a positive solution $X_{new} > 0$. It is almost common sense that we should have positivity of $(I - A)X > 0$, before computing the inverse in Eq. (8.20).

The necessary and sufficient condition for the existence of a positive solution $X_{new} > 0$ in Eq. (8.20) is known as the Hawkins-Simon condition. It states that the elements $(I - A) = \{b_{ij}\}$ should satisfy:

$$b_{11} > 0, \quad \begin{vmatrix} b_{11} & b_{12} \\ b_{21} & b_{22} \end{vmatrix} > 0, \quad \begin{vmatrix} b_{11} & b_{12} & \dots & b_{1n} \\ b_{21} & b_{22} & \dots & b_{2n} \\ \dots & \dots & \dots & \dots \\ b_{n1} & b_{n2} & \dots & b_{nn} \end{vmatrix} > 0. \tag{8.21}$$

where we have determinants of various 'principal minors' starting from top left corner of the $(I - A)$ matrix. They all need to to positive. Economists are familiar with similar conditions proposed by J. R. Hicks for stability of equilibrium. In the context of input-output analysis, Eq. (8.21) means that each industry must produce positive *net* output of its own goods. Takayama (1985) has references to the considerable economics literature on this topic.

Since prices cannot be negative, economists define a price vector as $p \geq 0$. Now, the amount of 'value added' per unit of output by j-th industry, v_j, can be defined as $v_j = p_j - \Sigma_{i=1}^{n} p_i a_{ij}$. It makes economic sense to require that $v_j > 0$ for each $j = 1, \dots n$. In matrix notation, collecting the n elements of v_j in vector v, we have the algebraic requirement that a price vector $p \geq 0$ exists such that:

$$v = (I - A)'p, \tag{8.22}$$

holds true. In fact, satisfaction of Eq. (8.21) also implies that the required price vector $p \geq 0$ exists to satisfy Eq. (8.22).

The 'Law of Demand' states that when quantity demanded increases for i-th good (beyond market-clearing equilibrium quantity) we have excess demand, and then the free market responds with a price rise for that good. In general, a (Walrasian) competitive trading system is mathematically described in terms of a numeraire price $p_0 = 1$ (or money) and nonlinear 'excess demand' functions $f_i(p)$ defined for each i-th good in the economy. Note that due to inter-dependence in the economy, excess demand functions are function of a vector of *all* n prices in the economy. When equilibrium price $\hat{p}$ vector is found there should be no excess demand, or excess demand should be zero: $f_i(\hat{p}) = 0$.

A dynamic competitive system must properly adjust to changing demand and supply conditions by satisfying $(dp/dt) = \dot{p} = f_i(p)$. A linear approximation will retain only the first order term in a Taylor series expansion. Section 15.5 discusses Taylor series in matrix notation. Here, it leads to $\dot{p} = A(p - \hat{p})$, where the matrix A contains partials of excess demand functions with respect to j-th price evaluated at equilibrium prices $p = \hat{p}$. The issue in equilibrium is whether prices tend to equilibrium prices as time passes, stated as $t \to \infty$. Matrix algebra helps in promoting a precise understanding of these issues and underlying conditions.

8.7.2 *Diagonal Dominance*

One of the interesting properties of the $n \times n$ input-output coefficient matrix A is that it must satisfy:

$$|a_{ii}| \geq \sum_{i \neq j, j=1}^{n} |a_{ij}|, \quad i = 1, \ldots, n, \tag{8.23}$$

where the vertical bars denote absolute values and ensure applicability of the definition to general matrices (beyond input-output coefficient matrices). Note that matrices satisfying Eq. (8.23) are known as 'Dominant diagonal' matrices, because their diagonal entries equal or exceed the row-sum of all off diagonal entries for each row. If the inequality is strict ($>$), the matrix is called 'Strictly dominant diagonal.' Gershgorin circle theorem states that such strictly (or irreducibly) diagonally dominant matrices are non-singular.

\# R program snippet **8.7.3** is next.
```
#Be sure to have earlier snippet in memory
F=diag(c(1.10, 1.15, 1.12)) %*% Finaldemand
F
Xnew=solve(diag(3)-A)%*% F; Xnew
```

The output of 8.7.3 below shows that all entries of X_{new} implied by prescribed growth percentages for all sectors are positive, implying that they make economic sense.

```
> F
       [,1]
[1,]   44.0
[2,] 149.5
[3,]  22.4
> Xnew=solve(diag(3)-A)%*% F; Xnew
          [,1]
[1,] 373.5205
[2,] 455.5365
[3,] 306.5975
```

Although our illustrative A is n×n, with $n = 3$ only, it is possible to use these techniques to determine detailed levels of thousands of outputs ($n >$ 1000) in terms of commodities and project employment in various industries and regions. The R snippets 8.7.1,2,3 work for very large matrices also subject to the memory of the computer.

In addition to policy modeling, an important contribution of input-output analysis is to provide logical arrangement of data into matrices to improve data accuracy. The reader who is interested in how to incorporate 'scrap' and other data adjustments and solutions to practical data problems is refereed to 14 mathematical equations in 'Appendix to Chapter 12' found in HP06 (available for free download on the Internet). It is interesting that all 14 equations involve matrix algebra and illustrate its practical use.

8.8   Partitioned Matrices

A matrix can be partitioned into sub-matrices at will. For example we have a partition when we define a minor of Eq. (6.2) by deleting a row and column of a matrix to end up with Eq. (6.3). For example, consider a partition of the original 4×4 matrix A into four 2×2 matrices:

$$A = \begin{bmatrix} a_{11} & a_{12} & a_{13} & a_{14} \\ a_{21} & a_{22} & a_{23} & a_{24} \\ a_{31} & a_{32} & a_{33} & a_{34} \\ a_{41} & a_{42} & a_{43} & a_{44} \end{bmatrix} = \begin{bmatrix} A_{11} & A_{12} \\ A_{21} & A_{22,} \end{bmatrix} \tag{8.24}$$

where A_{11} is formed from the top left corner 2×2 matrix. Now assume that we have a second 4×4 matrix B similarly partitioned.

8.8.1 Sum and Product of Partitioned Matrices

In general, let us define two $m \times n$ matrices

$$A = \begin{bmatrix} A_{11} & A_{12} \\ A_{21} & A_{22} \end{bmatrix}, \text{ and } B = \begin{bmatrix} B_{11} & B_{12} \\ B_{21} & B_{22} \end{bmatrix}, \tag{8.25}$$

where the submatrices have appropriate and comparable dimensions: A_{ij} is $m_i \times n_j$ for i and $j = 1, 2$. Now $m = m_1 + m_2, n = n_1 + n_2$. Such partitions of A and B are called conformable. Now we have the sum defined most naturally as:

$$A + B = \begin{bmatrix} A_{11} + B_{11} & A_{12} + B_{12} \\ A_{21} + B_{21} & A_{22} + B_{22} \end{bmatrix}. \tag{8.26}$$

The product AB is also naturally defined from the usual row-column multiplication explained in Sec. 4.4. We simply apply those rules to the blocks of partitioned matrices with subscripts on A and B in Eq. (8.24) similar to Eq. (8.26). That is, matrix multiplication problem involving two $m \times n$ matrices is simplified to multiplication of two 2×2 matrices. Of course this will not work unless all matrix multiplications are 'conformable' in the sense described in Sec. 4.4. Recall that the a matrix multiplication PQ is conformable only when the number of columns of P equals the number of rows of Q.

$$AB = \begin{bmatrix} A_{11}B_{11} + A_{12}B_{21} & A_{11}B_{12} + A_{12}B_{22} \\ A_{21}B_{11} + A_{22}B_{21} & A_{21}B_{12} + A_{22}B_{22} \end{bmatrix}. \tag{8.27}$$

8.8.2 Block Triangular Matrix and Partitioned Matrix Determinant and Inverse

The submatrices are called blocks and if $A_{12} = 0$ (the null matrix), A is called a lower block triangular matrix. If $A_{21} = 0$ it is called upper block triangular. If both $A_{12} = 0$ and $A_{21} = 0$ it is called a block diagonal matrix.

The determinant of a partitioned matrix is not intuitive, and needs to be learned. There are two answers for the above A of Eq. (8.24):

$$\begin{aligned} det(A) &= det(A_{11})det(A_{22} - A_{21}A_{11}^{-1}A_{12}) \\ &= det(A_{22})det(A_{11} - A_{12}A_{22}^{-1}A_{21}). \end{aligned} \tag{8.28}$$

This formula seems strange. However, it will seem less strange if we think of the submatrices as scalars, i.e., 1×1 matrices. Now write $det(A) =$

$A_{11}A_{22} - A_{21}A_{12}$ from the usual formula for a 2×2 determinant. Now take out the factor A_{11} and write the determinant as: $A_{11}\left(A_{22} - [A_{21}A_{12}]/A_{11}\right)$, and the formula Eq. (8.28) merely replaces the A_{11} in the denominator with A_{11}^{-1}. It is conformably placed in the middle of the expression to make sure that the matrix multiplication is well defined.

Now we use R to verify two matrix theory results on determinants of block diagonals and block triangular matrices:

If A is block diagonal, $det(A) = det(A_{11})det(A_{22})$, and furthermore that the same relation holds for upper or lower triangular A.

```
#R.snippet
set.seed(325); a=sample(10:200)
A=matrix(a[1:16],4); A#view Large AL
A11=matrix(c(a[1:2],a[5:6]),2);A11
A21=matrix(c(a[3:4],a[7:8]),2);A21
A12=matrix(c(a[9:10],a[13:14]),2);A12
A22=matrix(c(a[11:12],a[15:16]),2);A22
BlockDiag=A #initialize
BlockDiag[3:4,1:2]=0
BlockDiag[1:2,3:4]=0
BlockDiag #view BlockDiagonal matrix
LHS=det(BlockDiag);LHS
RHS=det(A11)*det(A22);RHS
round(LHS-RHS,6)#check determinants are equal
BlockTriang.UP=A #initialize
BlockTriang.UP[3:4,1:2]=0
BlockTriang.UP #view BlockTriang.UP matrix
LHS=det(BlockTriang.UP);LHS
RHS=det(A11)*det(A22);RHS
round(LHS-RHS,6)#check determinants are equal
BlockTriang.Lower=A #initialize
BlockTriang.Lower[1:2,3:4]=0
BlockTriang.Lower #view BlockTriang.Loweronal matrix
LHS=det(BlockTriang.Lower);LHS
RHS=det(A11)*det(A22);RHS
round(LHS-RHS,6)#check determinants are equal
```

We have chosen long self explanatory names to R objects. Product of determinants of diagonal blocks is computed by the command 'det(A11)*det(A22)'.

```
> A=matrix(a[1:16],4); A#view Large AL
     [,1] [,2] [,3] [,4]
[1,]   64   19  138   67
[2,]  157  108   71   95
[3,]   43   49  125  121
[4,]  198  196   61   93
> A11=matrix(c(a[1:2],a[5:6]),2);A11
     [,1] [,2]
[1,]   64   19
[2,]  157  108
> A21=matrix(c(a[3:4],a[7:8]),2);A21
     [,1] [,2]
[1,]   43   49
[2,]  198  196
> A12=matrix(c(a[9:10],a[13:14]),2);A12
     [,1] [,2]
[1,]  138   67
[2,]   71   95
> A22=matrix(c(a[11:12],a[15:16]),2);A22
     [,1] [,2]
[1,]  125  121
[2,]   61   93
> BlockDiag=A #initialize
> BlockDiag[3:4,1:2]=0
> BlockDiag[1:2,3:4]=0
> BlockDiag #view BlockDiagonal matrix
     [,1] [,2] [,3] [,4]
[1,]   64   19    0    0
[2,]  157  108    0    0
[3,]    0    0  125  121
[4,]    0    0   61   93
> LHS=det(BlockDiag);LHS
[1] 16674676
> RHS=det(A11)*det(A22);RHS
[1] 16674676
> round(LHS-RHS,6)#check determinants are equal
[1] 0
```

```
> BlockTriang.UP=A #initialize
> BlockTriang.UP[3:4,1:2]=0
> BlockTriang.UP #view BlockTriang.UP matrix
      [,1] [,2] [,3] [,4]
[1,]    64   19  138   67
[2,]   157  108   71   95
[3,]     0    0  125  121
[4,]     0    0   61   93
> LHS=det(BlockTriang.UP);LHS
[1] 16674676
> RHS=det(A11)*det(A22);RHS
[1] 16674676
> round(LHS-RHS,6)#check determinants are equal
[1] 0
> BlockTriang.Lower=A #initialize
> BlockTriang.Lower[1:2,3:4]=0
> BlockTriang.Lower #view BlockTriang.Loweronal matrix
      [,1] [,2] [,3] [,4]
[1,]    64   19    0    0
[2,]   157  108    0    0
[3,]    43   49  125  121
[4,]   198  196   61   93
> LHS=det(BlockTriang.Lower);LHS
[1] 16674676
> RHS=det(A11)*det(A22);RHS
[1] 16674676
> round(LHS-RHS,6)#check determinants are equal
[1] 0
```

The above snippet and its output are designed to show how the product of determinants of diagonal blocks equals the determinant of a matrix of diagonal blocks as well as upper and lower block triangular matrices. It is also intended to illustrate how to create and work with partitioned matrices in R.

The theory of partitioned matrices used to be valuable during the old days when computers were less available or could not handle large matrices. A great deal of ingenuity was needed to create the right (conformable) partition. The inverse of partitioned matrices using cofactors takes up considerable space in older matrix books. We need not learn to invert partitioned matrices in today's computing environment. For example, the 'solve' function of R to invert a matrix can be readily used for large matrices.

If partitioned matrix inverse is needed for theoretical discussion or for better understanding of the elements of A^{-1}, it is possible to use the following definitional relation $AA^{-1} = I$. Now we can use Eq. (8.27) wherein we simply replace B by A^{-1}. Now four terms on the right side of Eq. (8.27) and corresponding terms of the identity matrix provide four equations which are available for such deeper understanding of the elements of partitioned inverse. For example, the first equation is $I = A_{11}B_{11} + A_{12}B_{21}$.

Exercise 1.1 in Vinod and Ullah (1981) reports the following formula for partitioned inverse, intended to be useful in theoretical discussions. Upon defining $D = C - B'A^{-1}B$, the formula states:

$$\left(\begin{bmatrix} A & B \\ B' & C \end{bmatrix}\right)^{-1} = \begin{bmatrix} A^{-1} + A^{-1}BD^{-1}B'A^{-1} & -A^{-1}BD^{-1} \\ D^{-1}B'A^{-1} & D^{-1} \end{bmatrix}. \quad (8.29)$$

8.9 Applications in Statistics and Econometrics

An understanding of matrix algebra is beneficial in Statistics, Econometrics and many fields because it allows deeper understanding of the regression model. Consider the regression model Eq. (0.1) mentioned earlier in the preface,

$$y - X\beta + \varepsilon, \quad (8.30)$$

where y is a $T \times 1$ vector, X is $T \times p$ matrix, β is a $p \times 1$ vector and ε is $T \times 1$ vector. We generally assume that the expectation $E\varepsilon = 0$, is a null vector and the variance-covariane matrix is $\sigma^2 I_T$, or proportional to the $T \times T$ identity matrix I_T.

The ordinary least squares (OLS) estimator is obtained by minimizing the error sum of squares $ESS = \varepsilon'\varepsilon$ written as a function of the unknown β:

$$ESS = (y - X\beta)'(y - X\beta) = y'y - 2y'X\beta + \beta'X'X\beta \quad (8.31)$$

Note that we can use matrix calculus to differentiate ESS with respect to β. The partial of the first term $y'y$ is obviously zero, since it does not contain β. The other partials will be discussed later in Sec. 14.3 and Eq. (14.5). The first order (necessary) condition for a minimum is

$$0 = 2\beta'X'X - 2y'X. \quad (8.32)$$

The first order condition in Eq. (8.32) can be rewritten as $X'X\beta = X'y$ by transposing all terms. This system is known as 'normal equations' and was mentioned earlier. It involves p equations in the p unknowns in β and are known as 'normal equations.' We solve these equations for β to yield the ordinary least squares (OLS) estimator:

$$\hat{\beta}_{OLS} = (X'X)^{-1}X'y. \tag{8.33}$$

The assumption that the (variance-)covariance matrix is $\sigma^2 I_T$, is strong and can be readily relaxed if the covariance matrix denoted by Ω is known and invertible. Then, we first find the related matrix V^h defined as the reciprocal of the square root of Ω: $V^h = \Omega^{-1/2}$. It is obvious that V will be symmetric since covariance matrices are always symmetric, that is, $(V^h)\prime = V^h$. Thus, $(V^h)\prime V^h = \Omega^{-1}$. Now, we transform Eq. (8.30) as

$$V^h y = V^h X\beta + V^h\varepsilon \quad EV^h\varepsilon = 0, \quad EV^h\varepsilon\varepsilon'V^{h\prime} = V^h\Omega V^{h\prime} = I. \tag{8.34}$$

Since the transformation has made the covariance matrix proportional to the identity matrix, we can use the formula in Eq. (8.33) upon replacing y by $V^h y$ and X by $V^h X$ to yield what is called the generalized least squares (GLS) estimator:

$$\hat{\beta}_{GLS} = (X'(V^h)\prime V^h X)^{-1}X'(V^h)\prime V^h y, \tag{8.35}$$
$$= (X'\Omega^{-1}X)^{-1}X'\Omega^{-1}y,$$

where we use the definitional relation: $(V^h)\prime V^h = \Omega^{-1}$.

The variance-covariance matrix of the GLS estimator is

$$V(\hat{\beta}_{GLS}) = (X'\Omega^{-1}X)^{-1}. \tag{8.36}$$

If we use the ordinary least squares (OLS) despite knowing that $\Omega \neq I$ we have the following variance-covariance matrix:

$$V(\hat{\beta}_{OLS}) = E(X'X)^{-1}X'\epsilon\epsilon'X(X'X)^{-1} \tag{8.37}$$
$$= (X'X)^{-1}X'\Omega X(X'X)^{-1},$$

known as the sandwich. The middle matrix is the 'meat' and it is placed between two identical matrices representing the 'bread.'

If one uses OLS knowing that GLS is more appropriate, one pays a price in terms of efficiency of estimators. Vinod (2010) discusses some remedies for heteroscedasticity. Various measures of inefficiency can be defined in terms of the roots of the determinantal equation discussed by Rao and Rao (1998, Sec. 14.8):

$$|(X'X)^{-1}X'\Omega X(X'X)^{-1} - \theta(X'\Omega^{-1}X)^{-1}| = 0 \tag{8.38}$$

A simpler efficiency measure E_{ff} using the hat matrix $H = X(X'X)^{-1}X'$ by Bloomfield and Watson (1975) is based on the trace: $tr[H\Omega^2 H - (H\Omega H)(H\Omega H)]$. It is readily computed from the eigenvalues of Ω denoted as λ_i ordered from the largest to the smallest with respect to (wrt) i. It is given by

$$0 \le E_{ff} \le \frac{1}{4} \sum_{i=1}^{min(p,T)} (\lambda_i - \lambda_{T-i+1})^2. \qquad (8.39)$$

The result Eq. (8.39) is based on an application of the following Kantorovich inequality. Since it allows complex numbers it refers to 'Hermitian' matrices discussed in Sec. 11.2. In the regression context when the data are real numbers, the reader can think of Hermitian matrices as 'real-symmetric' matrices. If A is $T \times T$ Hermitian positive definite matrix with ordered eigenvalues λ_i, then for all nonzero vectors x,

$$1 \le \frac{x^*Ax}{x^*x} \frac{x^*A^{-1}x}{x^*x} \le \frac{(\lambda_1 + \lambda_T)^2}{4\lambda_1\lambda_T}. \qquad (8.40)$$

Sometimes reliable estimation of Ω is very difficult, and therefore researcher chooses to ignore the efficiency loss and focus instead on reliable inference for OLS regression coefficients. This requires reliable estimates of OLS standard errors, despite heteroscedasticity arising from $\Omega \ne I$. Upon substituting estimated $\hat{\Omega} = \Omega(\hat{\phi})$, (where ϕ represents estimates based on regression residuals) in Eq. (8.37) we have

$$V_{HAC}(\hat{\beta}_{GLS}) = (X'X)^{-1}(X'\hat{\Omega}X)(X'X)^{-1}, \qquad (8.41)$$

where the large $\hat{\Omega}$ matrix (not very reliably known to the researcher) enters Eq. (8.41) only through a small $p \times p$ matrix: $(X'\hat{\Omega}X)$. This revises the standard errors (in the denominators of usual t tests) to become heteroscedasticity and autocorrelation *consistent* (HAC) standard errors: $SE_{HAC}(b) = \sqrt{[diag(Cov_{HAC}(b))]}$.

8.9.1 *Estimation of Heteroscedastic Variances*

If $y_1, y_2, \ldots, y_n, (n > 2)$ are pairwise uncorrelated random variables, with a common mean μ but distinct heteroscedastic variances $\sigma_1^2, \sigma_2^2, \ldots \sigma_n^2$, respectively. Then, we have a linear model $y = X\mu + \epsilon$, where X is a column of ones.

The least squares estimator of μ is the sample mean $\bar{y} = (1/n)\Sigma_{i=1}^n y_i$ obtained by using the data $y_1, y_2, \ldots, y_n$. The residuals are estimated by

$\hat{\epsilon} = (y_i - \bar{y})$. The Hadamard product of residuals with themselves equals simply the vector of squared residuals. We estimate the sample variance as $s^2 = [1/(n-1)]\Sigma_{i=1}^n (y_i - \bar{y})^2$. The hat matrix when X is a column of ones is considered in Sec. 4.8 for item (4). This hat matrix becomes $H = (1/n)J_n$, as mentioned in Eq. (3.16) also. Now Hy provides deviations from the mean. If $M = I_n - H$, one can consider a system of linear equations

$$(M \odot M)\hat{\sigma}^2 = (\hat{\epsilon} \odot \hat{\epsilon}), \tag{8.42}$$

where $\hat{\sigma}^2$ is a vector of estimates of heteroscedastic variances. Then it can be shown, Rao and Rao (1998, p. 201), that inverse of Hadarmard products is the following analytically known matrix of a special form:

$$(M \odot M)^{-1} = \begin{bmatrix} a & b & \dots & b \\ b & a & \dots & b \\ . & . & . . . . . & . \\ b & b & \dots & a \end{bmatrix} \tag{8.43}$$

where $b = [-1/(n-1)(n-2)]$ and $a = b + n/(n-2)$. This is checked in the R snippet 8.9.1.1. After simplifications, an unbiased estimator of $\hat{\sigma}_1^2$, the first variance is $[n/(n-2)](y_1 - \bar{y})^2 - [1/(n-2)]s^2$.

```
# R program snippet 8.9.1.1 checks all off-diagonals are same.
library(matrixcalc)
M=diag(4)-(1/4)*matrix(1,4,4)
MM=hadamard.prod(M,M)
MMinv=solve(MM);MMinv
```

The snippet uses the Hadamard product defined later in Sec. 11.6 as implemented in the R package 'matrixcalc'.

```
> MMinv
           [,1]        [,2]        [,3]        [,4]
[1,]   1.8333333 -0.1666667 -0.1666667 -0.1666667
[2,]  -0.1666667  1.8333333 -0.1666667 -0.1666667
[3,]  -0.1666667 -0.1666667  1.8333333 -0.1666667
[4,]  -0.1666667 -0.1666667 -0.1666667  1.8333333
```

Observe that all off-diagonal elements equal $-1/6$ and diagonal elements equal $2 - 1/6$ as claimed in the formulas $b = [-1/(n-1)(n-2)]$ and $a = b + n/(n-2)$ with $n = 4$ from Eq. (8.43).

8.9.2 MINQUE Estimator of Heteroscedastic Variances

Rao and Rao (1998, Sec. 6.4) explain C. R. Rao's principle of minimum norm quadratic unbiased estimation (MINQUE) for estimation of heteroscedastic variances. Start with the Eq. (8.30) model: $y = X\beta + \varepsilon$, with a diagonal matrix V of variances $\{\sigma_i^2\}, i = 1, \ldots, T$. Now define a $T \times 1$ vector p of known constants p_i and a weighted sum of variances using these constants as coefficients. It can be defined by the matrix multiplication Vp. The MINQUE principle states that we seek a quadratic estimator $y'By$ of Vp such that the matrix B with elements b_{ij} satisfies the following conditions.

$$BX = 0, \tag{8.44}$$

$$\sum_{i=1}^{T} b_{ii}\sigma_i^2 = \sum_{i=1}^{T} p_i\sigma_i^2, \tag{8.45}$$

$$\| B \|^2 = \sum_{i=1}^{T}\sum_{j=1}^{T} b_{ij}^2 \tag{8.46}$$

Note that Eq. (8.44) implies that for all vectors β_0, the estimator $y'By$ of Vp is invariant in the sense that:

$$y'By = (y - X\beta_0)'B(y - X\beta_0). \tag{8.47}$$

The Eq. (8.45) implies that the estimator is unbiased. Equation (8.46) defines the norm of B to be minimized.

If the $(M \odot M)$ matrix of Eq. (8.42) is nonsingular (due to repeated measurements) the MINQUE of V is given by

$$\hat{V} = (M \odot M)^{-1}(\hat{\epsilon} \odot \hat{\epsilon}). \tag{8.48}$$

Many theoretical papers inspired by MINQUE are available under the heading 'variance components' models. An R package called 'maanova' by Hao Wu and others offers a function called 'fitmaanova' giving an option for MINQUE.

8.9.3 Simultaneous Equation Models

In many applications including econometrics only one regression equation does not properly describe the underlying relations. If a model has M simultaneous equations defined by

$$y_j = X_j\beta_j + \varepsilon_j, \quad j = 1, \ldots M, \tag{8.49}$$

where y_j contains observations on j-th set of endogenous (dependent) variables and X_j contains the j-th set of observations on p_j exogenous regressors along the columns. We can stack Eq. (8.49) into a very large system of 'seemingly unrelated regressions':

$$
\begin{bmatrix} y_1 \\ y_2 \\ \vdots \\ y_M \end{bmatrix} = \begin{bmatrix} X_1 & 0 & \dots & 0 \\ 0 & X_2 & \dots & 0 \\ \vdots & \vdots & \dots & \vdots \\ 0 & 0 & \dots & X_M \end{bmatrix} \begin{bmatrix} \beta_1 \\ \beta_2 \\ \vdots \\ \beta_M \end{bmatrix} \begin{bmatrix} \varepsilon_1 \\ \varepsilon_2 \\ \vdots \\ \varepsilon_M \end{bmatrix}. \tag{8.50}
$$

Let us denote the long vector of dimension $TM \times 1$ on the left hand side of Eq. (8.50) as $\tilde{y}$ and the large $TM \times \tilde{p}$ matrix as $\tilde{X}$, where $\tilde{p} = \Sigma_{j=1}^M p_j$ and where β_j is assumed to be $p_j \times 1$ vector of coefficients. Also denote the long error vector as $\tilde{\varepsilon}$. Now we have

$$
\tilde{y} = \tilde{X}\tilde{\beta} + \tilde{\varepsilon}. \tag{8.51}
$$

It is tempting to apply the OLS formula from Eq. (8.33) above to the large system of Eq. (8.51). However, such a choice will involve inverting a very large $\tilde{p} \times \tilde{p}$ matrix, $\tilde{X}'\tilde{X}$ which might not be numerically stable. If the error terms in various equations are independent of each other so that $E(\tilde{\varepsilon}\tilde{\varepsilon}') = \tilde{\sigma}^2 I_{TM}$ holds we can simply use OLS to estimate each j-th equation separately instead of working with a large system of M equations, and still get the same estimates of β_j.

Unfortunately, such independence is unrealistic for time series data (y_j, X_j), making contemporaneous errors (and endogenous variables) correlated across our M equations. However, if these correlations are known, it is possible to incorporate them in a GLS estimator Eq. (8.35), with the added bonus of improved efficiency of the estimator. Note that GLS requires knowledge of the large $TM \times TM$ covariance matrix $\tilde{\Omega}$.

$$
E\tilde{\varepsilon}\tilde{\varepsilon}' = \tilde{\Omega} = \begin{bmatrix} \sigma_{11}I & \sigma_{12}I & \dots & \sigma_{1M}I \\ \sigma_{21}I & \sigma_{22}I & \dots & \sigma_{2M}I \\ \vdots & \vdots & \vdots & \vdots \\ \sigma_{M1}I & \sigma_{M2}I & \dots & \sigma_{MM}I \end{bmatrix} = \Sigma \otimes I, \tag{8.52}
$$

where Σ is an $M \times M$ matrix with elements $\{\sigma_{ij}\}$. The last equation uses a compact Kronecker product notation described in detail later. We can readily implement the GLS provided we have an estimate of $\tilde{\Omega}^{-1}$. We shall see that result (vii) in Sec. 12.1 dealing with Kronecker products of matrices allows us to avoid inverting a large $MT \times MT$ matrix $\tilde{\Omega}$ directly. We can simply invert the small $M \times M$ matrix Σ and write the inverse needed by the GLS estimator as: $\Sigma^{-1} \otimes I$.

8.9.4 *Haavelmo Model in Matrices*

A simplified version of the Haavelmo macroeconomic model discussed in Vinod (2008a, Sec. 6.1) chapter 6 provides an example of the power of matrix algebra in working with a system of simultaneous equations in a typical econometric model. A simple matrix equation can describe a very complicated model with hundreds of equations.

The idea is to write the structural equations which come from macro economic theory by moving all errors to the right hand side as

$$Y\Gamma + XB = U, \tag{8.53}$$

where Y contains all M endogenous variables in the system ordered in some consistent manner into a $T \times M$ matrix, X contains all K exogenous variables in the system ordered in some suitable manner into a $T \times K$ matrix, the matrix Γ is $M \times M$ containing parameters associated with endogenous variables and B is $K \times M$ matrix of remaining parameters, and finally the matrix U is $T \times M$ which has errors for each of the M equations arranged side by side. Thus, we have rewritten the structural equations, which come from economic theory. Our aim is to estimate the Γ and B matrices, perhaps simultaneously.

The advantage of Eq. (8.53) is that the reduced form needed for its estimation can be written in general terms by using matrix algebra as follows. Assume that the Γ matrix is invertible, premultiply both sides of Eq. (8.53) by its inverse and write:

$$Y\Gamma\Gamma^{-1} + XB\Gamma^{-1} = U\Gamma^{-1}. \tag{8.54}$$

Now, moving endogenous variables to the left side, we can rewrite this as the *reduced form* of (6.1.17) as

$$Y = -XB\Gamma^{-1} + U\Gamma^{-1} = X\Pi + V, \tag{8.55}$$

where we have introduced a change of notation from $(-B\Gamma^{-1}) = \Pi$ and $U\Gamma^{-1} = V$. This is a general expression for the reduced form. The reduced form makes all variables on the right hand side exogenous and allows estimation similar to Eq. (8.50) or its GLS version based on Eq. (8.35) and Eq. (8.52).

8.9.5   *Population Growth Model from Demography*

In demography, matrix algebra is useful in the so-called Leslie (female) population growth model. The model begins with a set of initial populations $x_i^{(0)}$ from suitably divided k age groups. Let vector $x^{(t)}$ denote a $k \times 1$ vector representing populations at time t in all k age groups. The time unit is chosen so that each passage of one unit time transitions the population to the next age group.

In each time unit b_i daughters are born to the females x_i in the i-th age group. All these births add to the population in the first age group only. Also, s_i measures the proportion of females expected to survive and pass into the next age group. The deaths affect all age groups except the first. The Leslie model says that female population growth can be described by a recursive relation, best described using matrix algebra as:

$$x^{(t)} = Lx^{(t-1)}, \quad (t = 1, 2, \ldots), \tag{8.56}$$

where L denotes the Leslie matrix defined by

$$L = \begin{bmatrix} b_1 & b_2 & b_3 & \ldots & b_{k-1} & b_k \\ s_1 & 0 & 0 & \ldots & 0 & 0 \\ 0 & s_2 & 0 & \ldots & 0 & 0 \\ \vdots & \vdots & \vdots & \ldots & \vdots & \vdots \\ 0 & 0 & 0 & \ldots & s_{k-1} & 0 \end{bmatrix}, \tag{8.57}$$

where the births b_i affecting only first age group are represented by the first row of L, whereas survival proportions s_i affecting all remaining age groups appear along all remaining rows of the $k \times k$ matrix. The matrix L has only non-negative entries and has a special pattern.

Chapter 9

Eigenvalues and Eigenvectors

Eigenvalues and eigenvectors are so fundamental to all properties of matrices that we are tempted to discuss them immediately after matrices are defined. Their computation used to be difficult and it prevented their widespread use in matrix algebra texts of yesteryear. However, R allows such easy computation of eigenvalues that it is no longer necessary to relegate a discussion of these concepts to esoteric parts of matrix theory.

For example, we could not resist mentioning the seventh property of determinants in Sec. 6.2 that the determinant is a product of eigenvalues. An implication of this in Sec. 6.4 is that when one of the eigenvalues is zero the matrix is singular. The eighth in the list of properties of the rank of a matrix in Sec. 7.3 mentions that the rank equals the number of nonzero eigenvalues. The defining Eq. (7.11) for the trace involves sum of eigenvalues.

9.1 Characteristic Equation

Consider a square matrix A of dimension n. The scalar eigenvalue λ and an $n \times 1$ eigenvectors x of A are defined from a fundamental defining equation, sometimes called the characteristic equation.

$$A\,x = \lambda\,x, \tag{9.1}$$

where the eigenvector x must be further assumed to be nonzero. (why?)

The Eq. (9.1) obviously holds true when $x = 0$ with all n zero elements. The defining equation for eigenvalues and eigenvectors rules out the trivial solution $x = 0$. Since $x \neq 0$, something else must be zero for Eq. (9.1) to hold. It is not intuitively obvious, but true that the equation $Ax -$

$\lambda x = 0$ can hold when $(A x - \lambda I_n)x = 0$, where I_n denotes the identity matrix, which must be inserted so that matrix subtraction of λ from A is conformable (makes sense). Since $x \neq 0$, we must make the preceding matrix $(A x - \lambda I_n)$ somehow zero. We cannot make it a null matrix with all elements zero, because A is given to contain nonzero elements. However, we can make a scalar associated with the matrix zero. The determinant is a scalar associated with a matrix. Thus we require that the following determinantal equation holds.

$$det(A - \lambda I_n) = 0. \tag{9.2}$$

Let us use R to better understand the defining equation Eq. (9.1) by computing Eq. (9.2) for a simple example of a 2×2 matrix with elements 1 to 4. R computes the (first) eigenvalue to be $\lambda = 5.3722813$. The aim of the snippet is to check whether the determinant in Eq. (9.2) computed by the R function 'det' is indeed zero. The snippet 9.1.1 computes the Euclidean length of x defined in equation Eq. (3.3) of Chap. 3 to be unity, suggesting that vector x is on the 'unit circle' from the origin of a two-dimensional space.

```
# R program snippet 9.1.1 is next.
A=matrix(1:4,2);A #view A
ea=eigen(A) #object ea contains eigenvalues-vectors of A
lam=ea$val[1];lam #view lambda
x=ea$vec[,1];x # view x
det(A-lam*diag(2)) #should be zero
sqrt(sum(x^2)) #Euclidean length of x is 1
lam^2-5*lam #solving quadratic OK if this is 2
```

The last two lines in the snippet 9.1.1 are not clear at this point. We shall see that they check the length of x to be 1, and that $\lambda = 5.372281$ as a solution makes sense.

```
> A=matrix(1:4,2);A #view A
     [,1] [,2]
[1,]    1    3
[2,]    2    4
> ea=eigen(A) #object ea contains eigenvalues-vectors of A
> lam=ea$val[1];lam #view lambda
[1] 5.372281
```

```
> x=ea$vec[,1];x # view x
[1] -0.5657675 -0.8245648
> det(A-lam*diag(2)) #should be zero
[1] 0
> sqrt(sum(x^2)) #Euclidean length of x is 1
[1] 1
> lam^2-5*lam #solving quadratic
[1] 2
```

Thus in the example $\det(A - \lambda I_n) = 0$. More explicitly we have:

$$\text{If } A = \begin{bmatrix} 1 & 3 \\ 2 & 4 \end{bmatrix} \text{ then } (A - \lambda I_n) = \begin{bmatrix} 1-\lambda & 3 \\ 2 & 4-\lambda \end{bmatrix}. \tag{9.3}$$

The determinant can be written as $[(1 - \lambda)(4 - \lambda) - 2 * 3] = 0$, leading to a quadratic polynomial in λ as $[4 + \lambda^2 - \lambda - 4\lambda - 6] = 0$, with two roots, as in Eq. (2.4). With two roots it is obvious that we have two eigenvalues (characteristic roots).

Clearly, R computations claim that 5.372281 is one of the roots of $[\lambda^2 - 5\lambda - 2] = 0$. A direct check on a root is obtained by simply substituting $\lambda = 5.372281$ in $[\lambda^2 - 5\lambda - 2]$ and seeing that it equals zero. The last line of the R snippet 9.1.1 accomplishes this task of checking.

9.1.1 *Eigenvectors*

If the eigenvalue $\lambda = 5.372281$ satisfies the definition of Eq. (9.1), what does this fact imply about the elements of the eigenvector x?

$$(A - \lambda I_n)x = \begin{bmatrix} (1-\lambda)x_1 + 3x_2 \\ 2x_1 + (4-\lambda)x_2 \end{bmatrix} = Bx = \begin{bmatrix} 0 \\ 0 \end{bmatrix}. \tag{9.4}$$

which gives two equations in two unknowns elements x_1, x_2 of x upon substitution of the calculated eigenvalue $\lambda = 5.372281$. The first equation says $-4.372281x_1 + 3x_2 = 0$ and the second equation says $2x_1 - 1.372281x_2 = 0$.

We can attempt to solve this by the methods in Sec. 8.5 by computing the inverse matrix B^{-1}, except that $B^{-1}0$ is always zero and the solution $x_1 = 0 = x_2$ is not acceptable. Actually, the solution of a homogeneous system is not unique. This means that we can arbitrarily choose $x_1 = 1$ as a part of the solution, and need only determine the second part $x_2 = 4.372281/3 = 1.457427$. It is easy to verify that (Bx) with $x = (1, 1.457427)$ is one of those non-unique solutions, since

$(B\,x) = (0, 6.19013e - 07)$, which is the column vector (0,0) when rounded to 5 digits, consistent with Eq. (9.4).

The non-uniqueness problem is solved in matrix theory by insisting on the additional nonlinear requirement that the vector x lie on the unit circle, that is $||x|| = 1$, or its length be always 1, which is verified in the snippet 9.1.1 for the 2×1 eigenvector $x = (-0.5657675 - 0.8245648)$. For this choice $(B\,x) = (-1.828951e - 07, -2.665563e - 07)$, which is again (0,0) when rounded to 6 digits, consistent with Eq. (9.4).

9.1.2   *n Eigenvalues*

Note that in general our A is $n \times n$ matrix. When $n = 3$ Eq. (9.2) will be 3×3. Evaluating its determinant by the formula Eq. (6.4) of Chap. 6 we have a term of the type $(\lambda - a)\,(\lambda - b)\,(\lambda - c)$, with the largest power λ^3 being 3. Thus it will have 3 roots as eigenvalues also known as characteristic roots.

Now, let us recall the 'lessons' mentioned in the sequel of Eq. (2.4) in the current context and conclude that:

1] The number of roots of the determinantal equation Eq. (9.2) equals n, the highest power of λ. This means there are n eigenvalues $\lambda_i, i = 1, 2, \ldots, n$ when each root independently satisfies the defining equation Eq. (9.2). This shows how to find the n eigenvalues of A. We can define a diagonal matrix of eigenvalues:

$$\Lambda = \text{diag}\,(\lambda_i, i = 1, 2, \ldots, n). \tag{9.5}$$

2] The finding of roots (although non-trivial) is quite easy in R. Since polynomial roots can, in general, involve imaginary numbers, eigenvalues also can involve imaginary numbers.

9.1.3   *n Eigenvectors*

Given the $n \times n$ matrix A, we have seen that there are n eigenvalues λ_i as in Eq. (9.5). Associated with each eigenvalue, it seems intuitively plausible that there are n eigenvectors.

As stated above, matrix theory insists that all eigenvectors lie on the unit circle (be of Euclidean length 1). The R function 'eigen' provides all eigenvectors. For the 2×2 example of the earlier snippet, the two eigenvectors are available by issuing the R command: 'ea$vec.' as seen in the following output by R (assuming the R snippet 9.1.1 is in the memory of R).

```
> G=ea$vec; G #view matrix of eigenvectors
           [,1]        [,2]
[1,] -0.5657675 -0.9093767
[2,] -0.8245648  0.4159736
```

It is customary to collect all n eigenvectors into an $n \times n$ matrix G.

9.2 Eigenvalues and Eigenvectors of Correlation Matrix

Correlation coefficients r_{ij} are useful measures in Statistics to assess the strength of possible linear relation between x_i and x_j for pairs of variables. The correlation coefficients satisfy $-1 \leq \{r_{ij}\} \leq 1$, and the matrix A of correlation coefficients is always symmetric: $A' = A$, with ones along the diagonal. The snippet 9.2.1 creates a 10×3 matrix of x data and then uses the R functions 'cor' and 'rcorr' to create the 3×3 correlation matrix A.

```
# R program snippet 9.2.1 is next.
set.seed(93);x=matrix(sample(1:90),10,3)
A=cor(x);A#correlation matrix for columns of x
x[2,1]=NA#insert missing data in column 1
x[1,3]=NA#insert missing data in column 3
library(Hmisc) #bring Hmisc package into R memory
rcorr(x)#produces 3 matrices
```

Many times the data are missing for some variables and not for others. The R package 'Hmisc,' Frank E Harrell Jr and Others. (2010), computes the correlation coefficients for pairs of data using all available (non-missing) pairs and reports three matrices in the output below.

```
> set.seed(93);x=matrix(sample(1:90),10,3)
> A=cor(x);A#correlation matrix for columns of x
           [,1]        [,2]         [,3]
[1,] 1.0000000  0.18780373  0.30091289
[2,] 0.1878037  1.00000000 -0.03818316
[3,] 0.3009129 -0.03818316  1.00000000
> x[2,1]=NA#insert missing data in column 1
> x[1,3]=NA#insert missing data in column 3
> library(Hmisc) #bring Hmisc package into R memory
```

```
> rcorr(x)#produces 3 matrices
      [,1] [,2] [,3]
[1,] 1.00 0.25 0.34
[2,] 0.25 1.00 0.01
[3,] 0.34 0.01 1.00


n
      [,1] [,2] [,3]
[1,]   10    9    8
[2,]    9   10    9
[3,]    8    9   10


P
      [,1]    [,2]    [,3]
[1,]         0.5101 0.4167
[2,] 0.5101         0.9708
[3,] 0.4167 0.9708
```

The output of the R function 'rcorr' consists of **three matrices**. The first is the correlation matrix, same as the A matrix created by the R function 'cor.' The second matrix 'n' reports the number of non-missing data in each pair. The last output matrix 'P' reports the p-values for statistically testing the null hypothesis that the true correlation is zero. When p-values are less than 0.05, we reject the null hypothesis and conclude that the correlation is statistically significantly different from zero. Note that p-values are not defined for the unit correlation coefficient of a variable with itself. Hence the diagonals in the output matrix 'P' are left blank by the 'Hmisc' package.

The eigenvalues and eigenvectors of correlation matrices A are of interest in Statistics and provide a useful example for this chapter. Let us use only two variables and define the 2×2 correlation matrix with the correlation coefficient 0.7 as the off diagonal term.

$$\text{If } A = \begin{bmatrix} 1 & 0.7 \\ 0.7 & 1 \end{bmatrix} \text{ then } A - \lambda I_n = \begin{bmatrix} 1-\lambda & 0.7 \\ 0.7 & 1-\lambda \end{bmatrix} \tag{9.6}$$

Now the $det(A - \lambda I_n) = (1 - \lambda)^2 - 0.7^2$. It is easy to verify that its roots are 1.7 and 0.3 by using the R command 'polyroot(c(0.51,-2,1))' or by direct substitution. The matrix G of eigenvectors for this correlation matrix can be seen to be:

$$G = \begin{bmatrix} w & -w \\ w & w \end{bmatrix} \text{ where w} = 1/\sqrt{2} = 0.7071068. \tag{9.7}$$

In R this can be verified by the command 'eigen(matrix(c(1,.7,.7,1),2)).'

A remarkable result known as **Cayley-Hamilton** theorem states that the matrix A satisfies its own characteristic equation. This is easy to check when A is a correlation matrix. The characteristic equation is: $(0.51 - 2\lambda + \lambda^2 = 0$. Hence we need to verify whether $0.51\,I_2 - 2A + A^2$ is a null matrix with all zeros. Note that I_2, the identity matrix, is needed in the above expression. Otherwise, the polynomial for sum of matrices will not make sense. The checking is done in the following snippet.

```
# R program snippet 9.2.2 check Cayley-Hamilton theorem.
A=matrix(c(1,.7,.7,1),2); A
0.51*diag(2)-2*A +A%*%A
```

Note that 'A^2' will not work, since R simply squares each term. We need matrix multiplication of 'A' with itself.

```
> A=matrix(c(1,.7,.7,1),2); A
     [,1] [,2]
[1,]  1.0  0.7
[2,]  0.7  1.0
> 0.51*diag(2)-2*A +A%*%A
     [,1] [,2]
[1,]   0    0
[2,]   0    0
```

The above output of the snippet 9.2.2 shows that the characteristic polynomial indeed gives a 2×2 matrix of all zeros as claimed by the Cayley-Hamilton theorem.

9.3 Eigenvalue Properties

Some properties are collected here. 1] The (absolute values of) eigenvalues (characteristic roots) $\lambda_i(A)$ of A are ordered from the largest to the smallest as:

$$|\lambda_1| \geq |\lambda_2| \geq \ldots \geq |\lambda_n|. \tag{9.8}$$

2] The trace of A satisfies the following relation:

$$tr(A) = \sum_{i=1}^{n} a_{ii} = \sum_{i=1}^{n} \lambda_i(A). \tag{9.9}$$

3] The determinant of A satisfies the following relation:

$$det(A) = \prod_{i=1}^{n} \lambda_i(A) \tag{9.10}$$

4] Recalling the notation in Sec. 4.2 the matrix is usually assumed to contain real numbers: $A \in M_{m,n}(\Re)$. We explicitly specify $A \in M_{m,n}(\mathscr{C})$ only when complex numbers are admitted. A singular matrix is characterized by some eigenvalue being zero. Let iff denote if and only if and use the symbol $\forall$ to represent 'for all.'

If $A \in M_n(\mathscr{C})$, then A is nonsingular iff $\lambda_i(A) \neq 0$, $\forall i = 1, \ldots n$.   (9.11)

5] Consider an upper triangular n square matrix which admits complex numbers with components B and D as p square and q square matrices, respectively.

$$\text{If } A = \begin{bmatrix} B & E \\ 0 & D \end{bmatrix} \in M_n(\mathscr{C}) \text{ where } B \in M_p(\mathscr{C}), \ D \in M_q(\mathscr{C}) \tag{9.12}$$

then the n eigenvalues of A are simply the p eigenvalues of B and q eigenvalues of D. In other words, the zero block matrix below the diagonal and the block matrix E above the diagonal can be ignored for all practical purposes in computing the eigenvalues.

If we have an upper triangular matrix admitting complex numbers where all entries below the diagonal are zero, the eigenvalues are simply the entries on the main diagonal.

$$\text{If } A = \begin{bmatrix} a_{11} & & \ldots & a_{1n} \\ 0 & a_{22} & \ldots & a_{2n} \\ \vdots & & & \vdots \\ 0 & & \ldots & a_{nn} \end{bmatrix} \in M_n(\mathscr{C}), \text{ then } \lambda_i(A) = (a_{11}, \ldots, a_{nn}). (9.13)$$

6] If we add a constant times the identity, the eigenvalues also have that constant added.

If $c \in \mathscr{C}$, and $A \in M_n(\mathscr{C})$, then $\lambda_i(A + cI_n) = c + \lambda_i(A)$.   (9.14)

7] If we sandwich the matrix A between B^{-1} and B, (two bread matrices) the eigenvalues of A are unchanged.

If $B \in M_n(\mathscr{C})$, and nonsingular, $A \in M_n(\mathscr{C})$, then $\lambda_i(B^{-1}AB) = \lambda_i(A)$.
(9.15)

8] Let $\mathcal{N}$ denote a set of positive integers. If we raise a matrix to a power $p \in \mathcal{N}$ each eigenvalue is also raised to that power:

If $A \in M_n(\mathscr{C})$, and $p \in \mathcal{N}$ then $\lambda_i(A^p) = \lambda_i^p(A)$. (9.16)

9] If A is nonsingular so that A^{-1} is well defined, then its eigenvalues are reciprocals of the eigenvalues of A. More generally, the power p can be any negative integer. We have:

If $A \in M_n(\mathscr{C})$ is nonsingular, $p \in \mathcal{N}$ and $p \neq 0$, then $\lambda_i(A^p) = \lambda_i^p(A)$. (9.17)

10] If A^* is the complex adjoint of A then

$$\lambda_i(A^*) = \bar{\lambda}_i(A), i = 1, 2, \ldots n. \qquad (9.18)$$

9.4 Definite Matrices

There are four kinds of 'definite' matrices in the literature.

1] A matrix is said to be negative definite if all its eigenvalues are negative.
2] A matrix is said to be positive definite if all its eigenvalues are positive.
3] A matrix is said to be non-negative definite (positive semi-definite) if all its eigenvalues are positive or zero.
4] A matrix is said to be non-positive definite (negative semi-definite) if all its eigenvalues are negative or zero.

Consider a matrix X with all elements from the set of real numbers from $\Re$. The matrix theory claims the following results:

Result: A matrix defined as the product $X'X$ is non-negative definite. If X has full column rank, then $X'X$ is positive definite.

In the following snippet 9.4.1 we define 'X' to be a 5×3 matrix created from random numbers and then check the eigenvalues of $X'X$.

```
# R program snippet 9.4.1 is next.
set.seed(645); a=sample(20:100)
X=matrix(a[1:15],5,3);X #view X
xtx= t(X) %*%(X);xtx#view X'X
library(fBasics)
rk(xtx) # 3 columns means full column rank=3
ea=eigen(xtx) #create object ea
lam=ea$val;lam  #use $ to extract eigenvalues
min(lam) #smallest eigenvalue
```

In this example, X has full column rank of 3. The eigenvalues of $X'X$ are all positive. Hence it is positive definite.

```
> X=matrix(a[1:15],5,3);X #view X
     [,1] [,2] [,3]
[1,]   43   36   70
[2,]   74   26   94
[3,]   47   39   45
[4,]   58   53   99
[5,]   38   81   72
> xtx= t(X) %*%(X);xtx#view X'X
        [,1]  [,2]  [,3]
[1,] 14342 11457 20559
[2,] 11457 12863 17798
[3,] 20559 17798 30746
> rk(xtx) # 3 columns means full column rank=3
[1] 3
> ea=eigen(xtx) #create object ea
> lam=ea$val;lam  #use $ to extract eigenvalues
[1] 55245.1667  2398.2161   307.6172
> min(lam) #smallest eigen value
[1] 307.6172
```

9.5   Eigenvalue-eigenvector Decomposition

Note that we start with an $n \times n$ matrix $A = X'X$ with n^2 observable elements, compute n eigenvalues, place them all in a diagonal matrix Λ. in Eq. (9.5). Next, we collect all n eigenvectors into the matrix G, again with n^2 unknowns. Thus, it seems intuitively impossible that using only n^2 knowns in A we can find unique solutions for $n^2 + n$ unknowns in Λ, G together.

Recall that we have solved the non-uniqueness problem by imposing a seemingly arbitrary requirement that each eigenvector lie on the unit circle. Combining these results we have

$$X'X = A = G\Lambda G' \tag{9.19}$$

The following snippet builds on the earlier one to check that $A - G\Lambda G'$ is a matrix of all zeros, confirming Eq. (9.19). Assuming snippet 9.4.1 is in the memory of R we can issue the following commands.

R program snippet **9.5.1** is next.
```
Lam=diag(lam) #construct diagonal matrix
G=ea$vector;G #use $method and view matrix of eigenvectors
GLGt= G %*% Lam %*% t(G); GLGt
round(xtx-GLGt,10) #check that above eq. holds
```

Note that Eq. (9.19) decomposes the A matrix into a matrix multiplication of 3 matrices. The component matrices are available in R for further manipulation by using the dollar suffix to the R object 'ea' containing the eigenvalues and eigenvectors.

```
> G=ea$vector;G #use $method and view matrix of eigenvectors
             [,1]       [,2]        [,3]
[1,] -0.4985300  0.4266961   0.7545848
[2,] -0.4467376 -0.8724364   0.1981926
[3,] -0.7428952  0.2382965  -0.6255569
> GLGt= G %*% Lam %*% t(G); GLGt
        [,1]   [,2]   [,3]
[1,] 14342 11457 20559
[2,] 11457 12863 17798
[3,] 20559 17798 30746
> round(xtx-GLGt,10) #check that above eq. holds
        [,1] [,2] [,3]
[1,]      0    0    0
[2,]      0    0    0
[3,]      0    0    0
```

This verifies that left side of Eq. (9.19) minus the right side gives the zero for this illustration as claimed by matrix theory. The decomposition of Eq. (9.19) can also be written in terms of a sum of outer products of eigenvectors:

$$A = G\Lambda G' = \lambda_1 g_1 g_1' + \lambda_2 g_2 g_2' + \ldots \lambda_n g_n g_n' \qquad (9.20)$$

Certain related matrix algebra results are worth noting here. If x is an $n \times 1$ vector, the supremum of the ratio satisfies: $\sup[x'Ax]/[x'x] = \lambda_1$, and at the other end, the infimum of the ratio satisfies: $\inf[x'Ax]/[x'x] = \lambda_n$.

If B is an $n \times k$ matrix, the infimum over B of supremum over $B'x = 0$ of the ratio satisfies: $\inf_B \sup_{B'x=0}[x'Ax]/[x'x] = \lambda_{k+1}$; and at the other end, we have: $\sup_B \inf_{B'x=0}[x'Ax]/[x'x] = \lambda_{n-k}$. The limit λ_{k+1} is reached when $B = [g_1, g_2, \ldots g_k]$ consists of the designated columns of the matrix G of eigenvectors of A. The limit λ_{n-k} is reached when $B = [g_{n-k+1}, g_{n-k+2}, \ldots g_n]$ similarly comprising the designated eigenvectors of A.

See Rao and Rao (1998, Ch. 10) and Rao (1973, sec 11f.2) for additional results including Sturmian and Poincare separation theorems involving sequences of matrices, matrix quadratic forms and matrix eigenvalues. We provide an example here.

The **Poincare Separation Theorem for Eigenvalues** is as follows: Let $A \in M_n$ be a Hermitian matrix defined later in Chapter 11. If all entries are real numbers we can think of the Hermitian matrix as a real symmetric matrix. Let the eigenvalues of A satisfy the inequalities: $\lambda_1 \geq \lambda_2 \geq \lambda_n$. Also let B be any matrix of order $n \times k$ such that columns of B form an orthonormal (unit norm orthogonal) set: $B^*B = I_k$. Now the eigenvalues $\beta_1 \geq \beta_2 \geq \beta_k$ of B^*AB are interlaced between those of A as:

$$\lambda_i \geq \beta_i \geq \lambda_{i+n-k}, \quad i = 1, 2, \ldots, k. \tag{9.21}$$

9.5.1 *Orthogonal Matrix*

We have defined and plotted angles between vectors in Chap. 3. If the angle is 90 degrees by Eq. (3.4) or Eq. (3.6) the vectors are orthogonal or perpendicular.

Matrix theory states that the G matrix of eigenvectors based on a positive definite matrix is such that: (a) each column vector of G lies on the unit circle, and (b) G is orthogonal, in the sense that its transpose equals its inverse.

Let us verify the orthogonality property stated as:

$$G'G = I_n = GG' \tag{9.22}$$

in the following snippet.

```
# R program snippet 9.5.1.1 Orthogonal Matrix.
ggt= G %*% t(G);
round(ggt,10)# view GG'
gtg= t(G) %*% G;
round(gtg,10) #view G'G
invG=solve(G);invG #G inverse
trG=t(G)#transpose of G
round(invG-trG,10)
```

As the following output shows G satisfies Eq. (9.22).

```
> round(ggt,10)# view GG'
     [,1] [,2] [,3]
[1,]   1    0    0
[2,]   0    1    0
[3,]   0    0    1
> gtg= t(G) %*% G;
> round(gtg,10) #view G'G
     [,1] [,2] [,3]
[1,]   1    0    0
[2,]   0    1    0
[3,]   0    0    1
> round(invG-trG,10)
     [,1] [,2] [,3]
[1,]   0    0    0
[2,]   0    0    0
[3,]   0    0    0
```

Toward the end of the snippet 9.5.1.1 we verify that G is is indeed an orthogonal matrix in the sense that its inverse equals its transpose by numerically checking that: $G^{-1} - G' = 0$.

The determinant of orthogonal matrix G is either +1 or -1. To make these results plausible, recall that the first two properties from Sec. 6.2 containing properties of determinants state that:

1] $det(A) = det(A')$, and

2] $det(AB) = det(A)det(B)$.

Hence applying them to Eq. (9.22) we know that $det(G) = det(G')$ and $det(G\,G') = det(I_n) = 1$, where we use the result that the determinant of the identity matrix is always unity. Thus, $det(G) = 1 = det(G')$ holds.

The eigenvalues of G are also +1 or -1. If an orthogonal vector is of unit norm, it is called 'orthonormal' vector.

A remarkable concept called Gram-Schmidt orthogonalization discussed in Sec. 17.4 is a process which assures the following. Any given n linearly independent vectors can be transformed into a set of n orthonormal vectors by a set of linear equations. If we start with a matrix A of full column rank and apply the Gram-Schmidt orthogonalization to its column vectors we end up with its QR decomposition $A = QR$, where Q is orthogonal and R is a triangular matrix. The QR decomposition discussed in Sec. 17.4.1 uses

similar methods and is available in R as a function 'qr.' Some details are
available in R manuals. Also see:

  http://en.wikipedia.org/wiki/Gram%E2%80%93Schmidt_process

9.6   Idempotent Matrices

In ordinary algebra the familiar numbers whose square is itself are: 1 or 0.
Any integer power of 1 is 1. In matrix algebra, the Identity matrix I plays
the role of the number 1, and I certainly has the property $I^n = I$.

There are other non-null matrices which are not identity and yet have
this property, namely $A^2 = A$. Hence a new name "idempotent" is needed
to describe them. These are usually symmetric matrices also.

Consider the so-called hat matrix defined as:

$$H = X(X'X)^{-1}X'. \tag{9.23}$$

Equation (9.25) below explains placing a hat on y through this matrix.

It is easy to verify that H is symmetric ($H' = H$) and recall from
Eq. (4.14) that it is idempotent: $H^2 = H$, because of a cancellation of the
identity matrix in the middle:

$$H^2 = X(X'X)^{-1}X'X\,(X'X)^{-1}X' = XI(X'X)^{-1}X' = H \tag{9.24}$$

We can extend Eq. (9.24) to show $H^n = H$. This has important implications for eigenvalues $\lambda_i(H)$ of the idempotent matrix H, namely that the
eigenvalues are either 1 or 0. The hat matrix H can be viewed as a projection matrix, where the geometric name comes from viewing the matrix
multiplication

$$Hy = X(X'X)^{-1}X'y = X\hat{\beta} = \hat{y}, \tag{9.25}$$

as placing a hat (representing fitted value in the standard notation) on y.
Note that in Eq. (9.25) we replace a part of the expression by the ordinary
least squares (OLS) estimator in the usual regression model $y = X\beta + \varepsilon$,
discussed later in Sec. 12.2 in Eq. (12.5). It is thought of as a geometrical
projection of the y vector on the regression hyperplane based on the column
space of the $T \times p$ matrix of regressors X.

Another projection matrix from regression theory is $M = I_T - H$ which
projects the y vector on the space of residuals to yield a column vector of
observable regression residuals $\hat{\varepsilon}$.

$$My = (I_T - H)y = (I_T - X(X'X)^{-1}X')y = y - \hat{y} = \hat{\varepsilon}. \qquad (9.26)$$

We claimed at the beginning of this chapter that eigenvalues are fundamental for a matrix. Note that the idempotent matrices are fundamentally
similar to the identity matrix since its square is itself.

Now we list some properties of the idempotent matrices:

(1) Since the eigenvalues of idempotent matrices are 1 or 0, they behave
like the identity (or null) matrix in a certain sense.

(2) First, partition the $T \times p$ matrix $X = (X_1, X_2)$ and define the hat matrix
for X_1 as $H_1 = X_1(X_1'X_1)^{-1}X_1'$. Now we define three matrices:$A_1 =
I - H$, $A_2 = H_1$, and $A_3 = H - H_1$. Note that they satisfy the identity:
$A_1 + A_2 + A_3 = I - H + H_1 + H - H_1 = I$.

Given a random $T \times 1$ vector y, this property implies a result in
'quadratic forms' discussed in Chapter 10 useful in Statistical technique
called 'analysis of variance':

$$y'y = y'A_1y + y'A_2y + y'A_3y. \qquad (9.27)$$

(3) If A is symmetric idempotent, then $C = I - 2A$ is orthogonal because
the transpose of C equals its inverse, that is $CC' = I$.

(4) If two matrices are idempotent, their Kronecker product is also idempotent.

(5) If A, B are $n \times n$, then AB, BA are idempotent if either $ABA = A$ or
$BAB = B$.

(6) If A, B are $n \times n$ idempotent matrices, then AB, BA are also idempotent
if they commute in the sense that $AB = BA$.

(7) If A, B are $n \times n$ symmetric idempotent matrices, then $A - B$ is also
symmetric idempotent if and only if $B^2 = BA$.

(8) Any symmetric $n \times n$ matrix A can be decomposed into a weighted
sum of k disjoint symmetric idempotent matrices; $A = \Sigma_{i=1}^k d_i A_i$, where
disjoint means $A_i A_j = 0$ for all $i \neq j$.

(9) The trace of an idempotent matrix equals its rank. This follows from
three results that (a) eigenvalues are 1 or 0, (b) trace equals the sum
of eigenvalues, and (c) rank is the number of nonzero eigenvalues.

(10) If $A_i, (i = 1, 2, \ldots k)$ are symmetric matrices of rank n_i such that
$\Sigma_{i=1}^k A_i = I$ and also $\Sigma_{i=1}^k n_i = n$, then $A_i A_j = 0, (i \neq j = 1, 2, \ldots k$
(disjointness) and $A_i = A_i^2$ being idempotent.

\# R program snippet **9.6.1** Hat Matrix is Idempotent.

```
library(fBasics) #so we can use rk()
set.seed(645); a=sample(20:100)
X=matrix(a[1:15],5,3)#view X
xtx= t(X) %*%(X);xtx#view X'X
H=X %*% solve(xtx) %*% t(X)
round(H,4) #view the hat matrix
sum(diag(H)) #trace by direct method
d=eigen(H)$values
round(d,5) #View eigenvalues of H
rk(H) #rank of H
sum(d) #sum of eigenvalues of H
M=diag(5)-H; 5 by 5 I-Hat is M matrix
round(M,4) #view the hat matrix
sum(diag(M)) #trace of M by direct method
dm=eigen(M)$values # eigenvalues of M
round(dm,5) #eigenvalues of M
rk(M)#rank of M
sum(dm) #sum of eigenvalues of M
```

The reader should remember to pay attention to my comments in the snippet 9.6.1, as before, so that its output below is easy to understand.

```
> library(fBasics) #so we can use rk()
> set.seed(645); a=sample(20:100)
> X=matrix(a[1:15],5,3)#view X
> xtx= t(X) %*%(X);xtx#view X'X
       [,1]   [,2]   [,3]
[1,] 14342 11457 20559
[2,] 11457 12863 17798
[3,] 20559 17798 30746
> H=X %*% solve(xtx) %*% t(X)
> round(H,4) #view the hat matrix
        [,1]     [,2]     [,3]     [,4]     [,5]
[1,]  0.2081   0.2091  -0.0903   0.3123   0.1240
[2,]  0.2091   0.6774   0.2239   0.2431  -0.2562
[3,] -0.0903   0.2239   0.8401  -0.2079   0.1813
[4,]  0.3123   0.2431  -0.2079   0.4806   0.2231
[5,]  0.1240  -0.2562   0.1813   0.2231   0.7939
> sum(diag(H)) #trace by direct method
[1] 3
```

```
> d=eigen(H)$values
> round(d,5) #View eigenvalues of H
[1] 1 1 1 0 0
> rk(H) #rank of H
[1] 3
> sum(d) #sum of eigenvalues of H
[1] 3
> M=diag(5)-H; 5 by 5 I-Hat is M matrix
Error: unexpected symbol in " 5 by"
> round(M,4) #view the hat matrix
        [,1]      [,2]      [,3]      [,4]      [,5]
[1,]   0.7919  -0.2091   0.0903  -0.3123  -0.1240
[2,]  -0.2091   0.3226  -0.2239  -0.2431   0.2562
[3,]   0.0903  -0.2239   0.1599   0.2079  -0.1813
[4,]  -0.3123  -0.2431   0.2079   0.5194  -0.2231
[5,]  -0.1240   0.2562  -0.1813  -0.2231   0.2061
> sum(diag(M)) #trace of M by direct method
[1] 2
> dm=eigen(M)$values # eigenvalues of M
> round(dm,5) #eigenvalues of M
[1] 1 1 0 0 0
> rk(M)#rank of M
[1] 2
> sum(dm) #sum of eigenvalues of M
[1] 2
```

Thus the claims made following Eq. (9.26) are all verified for this artificial example of random data with three regressors.

Note that the projection M is such that post-multiplication by X gives the null matrix (all zeros).

$$MX = (I - H)X = X - HX = X - X = 0. \qquad (9.28)$$

Now Eq. (9.26) can be rewritten as

$$\hat{\varepsilon} = My = M(X\beta + \varepsilon) = M\varepsilon, \qquad (9.29)$$

which relates the observable residuals to the unobservable true unknown errors of the regression model. This is an important result in regression theory.

The observable 'residual sum of squares' is written as:

$$(\hat{\varepsilon}'\hat{\varepsilon}) = (\varepsilon'M'M\varepsilon), \qquad (9.30)$$

where $M'M = M^2 = M$, since M is a symmetric idempotent matrix. If true unknown regression errors ε are assumed to be Normally distributed, their sum of squares is a Chi-square random variable discussed further in Sec. 15.1. The observable $(\hat{\varepsilon}'\hat{\varepsilon})$ is related to a quadratic form $(\varepsilon'M\varepsilon)$, based on Eq. (9.30), (to be discussed in detail in Sec. 10.1). Hence it can be shown that $(\hat{\varepsilon}'\hat{\varepsilon}) \sim \chi^2_{T-p}$, a Chi-square variable with degrees of freedom $(T - p)$ equal to the trace of the symmetric idempotent (projection) matrix M: trace$(M) = T - p$, when X is a $T \times p$ matrix. An implication is that the degrees of freedom of the familiar t-test of regression coefficients depend on the trace of a symmetric idempotent matrix. Thus, matrix algebra results regarding the trace have important practical implications in Statistics.

9.7 Nilpotent and Tripotent matrices

If $A^k = 0$ for some integer k, it is called a Nilpotent matrix.

If an $n \times n$ matrix A satisfies $A = A^3$, it is called tripotent, an extension of idempotent concept to the third power.

Now we list some properties of the tripotent matrices A:

(1) If P is orthogonal $n \times n$ matrix, $P'AP$ is also tripotent. If A is symmetric tripotent then $P'AP$ is symmetric idempotent.
(2) Pre-multiplication and post-multiplication of A by a non-singular matrix P retains tripotent property of $P^{-1}AP$.
(3) Both $-A$ and A^2 retain the (symmetric) tripotent property of A.
(4) The eigenvalues of A are $-1, 0$ or 1. In fact, any symmetric matrix is symmetric tripotent if and only if its eigenvalues are $-1, 0$ or 1.
(5) The rank of A equals the trace $tr(A^2)$.
(6) A necessary and sufficient condition for a symmetric matrix B to be tripotent is that B^2 is idempotent.
(7) If A is nonsingular tripotent, $A^{-1} = A, A^2 = I$ and $(A+I)(A-I) = 0$.

Chapter 10

Similar Matrices, Quadratic and Jordan Canonical Forms

We have noted in Eq. (9.15) of Chap. 9 that a certain 'sandwich' matrix has same eigenvalues. Also recall Eq. (9.19) definition of eigenvalue-eigenvector decomposition where the diagonal matrix of eigenvalues is present. In this chapter we consider the concept of diagonalizability of a square matrix and related concepts involving Quadratic forms and Jordan matrices. We begin with Quadratic forms.

10.1 Quadratic Forms Implying Maxima and Minima

A quadratic function in one variable $Q(x) = ax^2$ is sometimes called a quadratic form. If $a > 0$ the function $Q(x) = ax^2 \geq 0$ and it equals zero only when $x = 0$. That is, we have a useful practical result that $x = 0$ globally minimizes the function $Q(x)$ which can be described as a 'positive definite' quadratic form. If $a < 0$, we have a 'negative definite' quadratic form $Q(x)$ which is globally maximized at $x = 0$.

In two dimensions, the form $Q_1(x, y) = x^2 + y^2 > 0$ is a quadratic form similar to Q above and equals zero only when $x = y = 0$. Again, the global minimum of Q_1 is at $x = y = 0$. Similarly, the form $Q_2(x, y) = -x^2 - y^2 < 0$ is a negative definite quadratic form equals zero only when $x = y = 0$. Again, the global maximum of Q_2 is at $x = y = 0$.

However the third form $Q_3(x, y) = x^2 - y^2$ is a quadratic form with an ambiguous sign in the sense that $Q_3 > 0$ when $y = 0$, but $Q_3 < 0$ when $x = 0$. Since it can be positive or negative, it is called 'indefinite quadratic form'.

Now the fourth form, $Q_4(x, y) = (x+y)^2 > 0$, is a 'non-negative definite' or 'positive semidefinite'. It is zero when $x = y = 0$, but Q_4 is also zero whenever $x = -y$. Hence $x = y = 0$ is not a unique global minimizer of the quadratic form. The analogous fifth form $Q_5(x, y) = -(x + y)^2 < 0$, is a 'non-positive definite' or 'negative semidefinite'. It is zero when $x = y = 0$, but it is also zero whenever $x = -y$. Again the maximizer is not unique.

It is fun to use the three dimensional plotting tools in R to visualize the shapes Q_1 to Q_4 defined above. We omit the Q_5 plot for brevity. The reader may wonder what this has to do with matrices. We shall shortly define in Eq. (10.1) a 2×1 vector $x = (x_1, x_2)$ instead of (x, y), and write the quadratic function as a row-column matrix multiplication obtained from the product $x'Ax$, with a suitable choice of the matrix A in the middle. The notation (x, y) is more convenient for plotting purposes where we choose a range of x, y values along the two axes in the closed interval $[-3, 3]$ evaluated with increments of 0.2, implemented by the R command 'x=seq(-3,3,by=.2)'.

```
# R program snippet 10.1.1 plots quadratic forms.
par(mfrow=c(2,2))   #four plots in one
x=seq(-3,3,by=.2)
y=seq(-3,3,by=.2)
f=function(x,y){x^2+y^2}
z=outer(x,y,f)
persp(x,y,z, main="Positive definite Quadratic Form")
#plots the surface
#second plot
f=function(x,y){-x^2-y^2}
z=outer(x,y,f)
persp(x,y,z, main="Negative definite Q. Form")
#third plot
f=function(x,y){x^2-y^2}
z=outer(x,y,f)
persp(x,y,z, main="Indefinite Quadratic Form")
#fourth plot
f=function(x,y){(x+y)^2}
z=outer(x,y,f)
persp(x,y,z,main="Non-negative Definite Q. Form")
par(mfrow=c(1,1))   #reset plot parameters
```

The R command 'outer' evaluates the function at all values of x, y in the designated interval and computes the heights of the curve for plotting. The function 'persp' creates a 3D or perspective plot. The output of snippet 10.1.1 is Figure 10.1 with four plots for functions Q_1 to Q_4 defined above.

Positive definite Quadratic Form   **Negative definite Q. Form**

Indefinite Quadratic Form   **Non–negative Definite Q. Form**

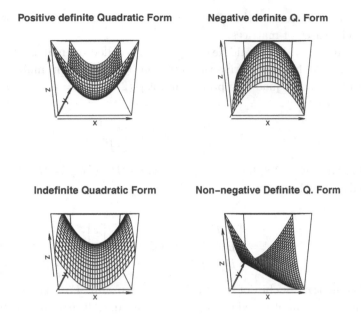

Fig. 10.1   Three dimensional plots of four quadratic form functions defined by functions Q_1 to Q_4.

If x is an $n \times 1$ vector and A is an $n \times n$ matrix the expression $x'Ax$ is of dimension 1×1 or a scalar. A scalar does not mean that it is just one term in an expression, only that when it is evaluated it is a number. For example, when $n = 2$, and $A = \{a_{ij}\}$ note that:

$$x'Ax = (x_1, x_2) \begin{bmatrix} a_{11} & a_{12} \\ a_{21} & a_{22} \end{bmatrix} \begin{bmatrix} x_1 \\ x_2 \end{bmatrix} \tag{10.1}$$

$$= (a_{11}x_1^2 + a_{12}x_1x_2 + a_{21}x_2x_1 + a_{22}x_2^2),$$

where the last expression evaluated by the usual row-column multiplication is a quadratic polynomial. Hence $x'Ax$ is called a quadratic form, which is a scalar and has an expression having 4 terms. Note that the largest power of any element of x is 2 even for the case when A is a 3×3 matrix. When the a_{ij} and x_i are replaced by numbers, it remains true that the quadratic form is just a scalar number, distinguished from a matrix. While a scalar cannot be a matrix, a matrix can be a scalar when its dimension is 1×1.

However, we have seen that R software rules of syntax do not allow scalars as special cases of matrices.

The matrix A in the middle of Eq. (10.1) is called the matrix of quadratic form $x'Ax$. It helps to derive the equation by row-column multiplications. Starting with the quadratic polynomial Eq. (2.4) $f(x) = a\,x^2 + b\,x + c = 0$, the matrix is

$$A = \begin{bmatrix} a & 0 \\ b & 1 \end{bmatrix}, \text{ and } x = \begin{bmatrix} x \\ 1 \end{bmatrix}. \tag{10.2}$$

Similarly, Eq. (2.8) function $f(x) = Ax^2 + 2Bxy + Cy^2 + 2Dx + 2Ey + F = 0$ needs a 3×3 matrix

$$A = \begin{bmatrix} A & B & D \\ B & C & E \\ D & E & F \end{bmatrix} \quad \text{and} \quad x = \begin{bmatrix} x \\ y \\ 1 \end{bmatrix}. \tag{10.3}$$

The matrix algebra of quadratic forms is useful. For example, one can use the determinant based test $det(A) = 0$ in Eq. (2.9) to help conclude that the function $f(x)$ represents two straight lines instead of only one curve. Section 11.8 discusses maxima and minima for ratios of quadratic forms used in Statistical hypothesis testing.

10.1.1 *Positive, Negative and Other Definite Quadratic Forms*

Just as a number can be negative, zero or positive, a quadratic form also can be negative, zero or positive. Since the sign of zero can be negative or positive, it is customary in matrix algebra to say that a quadratic form is negative or positive definite, meaning that $x'Ax < 0$ and $x'Ax > 0$, respectively. The expression "definite" already used above reminds us that it is not zero. If $x'Ax \geq 0$, it is called non-negative definite (nnd); and if $x'Ax \leq 0$, it is called non-positive definite (npd).

Note from Eq. (10.2) that the matrix A must be a square matrix for it to be the matrix of a quadratic form, but it need not be a symmetric matrix (i.e. $A' = A$ is unnecessary). If one has a symmetric matrix, the cross product terms can be merged. For example, the symmetry of A means that $a_{12} = a_{21}$ in the above 2×2 illustration:

$$a_{11}x_1^2 + a_{12}x_1x_2 + a_{21}x_2x_1 + a_{22}x_2^2 = a_{11}x_1^2 + 2a_{12}x_1x_2 + a_{22}x_2^2. \tag{10.4}$$

Suppose that one wishes to simplify the quadratic form as on the right side of Eq. (10.4), even when the matrix A is asymmetric. It is interesting to note that we can replace the matrix A of the quadratic form by $B = (1/2)(A + A')$, and use $x'Bx$ as the new quadratic form, which can be proved to be equal to $x'Ax$.

How to determine whether a given matrix of a quadratic form is positive definite or non-negative definite?

In terms of determinants, there is a sequence of determinants which should be checked to be positive definite for the quadratic form matrix to be positive definite. Economists refer to it as the Hawkins-Simon condition in Eq. (8.21) when we check:

$$a_{11} > 0, \quad \begin{vmatrix} a_{11} & a_{12} \\ a_{21} & a_{22} \end{vmatrix} > 0, \quad \begin{vmatrix} a_{11} & a_{12} & a_{13} \\ a_{21} & a_{22} & a_{23} \\ a_{31} & a_{32} & a_{33} \end{vmatrix} > 0, \tag{10.5}$$

where we have indicated the principal determinants of orders $= 1, 2, 3$. To see the intuitive reason for these relations consider $n = 2$ and the quadratic form

$$Q = a_{11}x_1^2 + 2a_{12}x_1x_2 + a_{22}x_2^2 \tag{10.6}$$

Complete the square by adding and subtracting $(a_{12}/a_{11})^2 x_2^2$. Now write Eq. (10.6) as:

$$Q = a_{11}[x_1 + (a_{12}/a_{11})x_2]^2 + [a_{22} - (a_{12}^2/a_{11})]x_2^2. \tag{10.7}$$

Observe that $Q > 0$ requires that $a_{11} > 0$ from the first term, and $a_{11}a_{22} - a_{12}^2 > 0$ from the second term. The second term is positive if and only if the 2×2 determinant above is positive. This explains the intuition.

Instead of computing cumbersome principal minors and their determinants, the question whether the matrix A in the middle of a quadratic form is positive definite is easily checked in the modern era of computers as follows. We simply compute the eigenvalues of A, and concentrate on the smallest eigenvalue. If the smallest eigenvalue satisfies: $\lambda_{min}(A) > 0$, is strictly positive, then A is said to be positive definite. Similarly, if the smallest eigenvalue $\lambda_{min}(A) \geq 0$, which can be zero, then the middle matrix A is said to be non-negative definite (nnd). Negative definite and non-positive definite (npd) quadratic forms can be analogously defined.

10.2   Constrained Optimization and Bordered Matrices

Recall from Sec. 10.1 that quadratic form functions $Q, Q_1, \ldots Q_4$ can help determine the global uniqueness of their maxima or minima, depending on whether the quadratic form is positive (negative, non-positive, non-negative) definite. Changing the notation from (x, y) to the vector $x = (x_1, x_2)$ it is easy to verify that a unique global minima or maxima is $x_1 = 0 = x_2$, only when the quadratic form $x'Ax$ is positive or negative definite. This holds true only if the maximization or minimization is not subject to any (linear) constraints. In practical applications constrained optimization is very common. Hence it is interesting to use matrix algebra to gains insights for constrained optimization problems.

Consider optimization of a general quadratic form in two variables $Q(x_1, x_2)$ defined as:

$$Q(x_1, x_2) = ax_1^2 + 2bx_1x_2 + cx_2^2 = (x_1, x_2) \begin{bmatrix} a & b \\ b & c \end{bmatrix} \begin{bmatrix} x_1 \\ x_2 \end{bmatrix}, \qquad (10.8)$$

subject to the linear constraint implying a linear subspace:

$$Ux_1 + Vx_2 = 0. \qquad (10.9)$$

The optimization problem can be solved by using Eq. (10.9) to eliminate one of the variables, say x_1 by writing it as $x_1 = -Vx_2/U$. Substituting this in the quadratic function (objective function Q) verify that it becomes:

$$Q = [(aV^2 - 2bUV + cU^2)/U^2] x_2^2. \qquad (10.10)$$

The sign of the bracketed coefficient determines the definiteness of the quadratic form Q subject to the linear constraint. The sign of the bracketed term in Eq. (10.10) is the same as the sign of its numerator, because the denominator is always positive. Since the numerator can be written as a negative determinant of a bordered matrix, we have the following useful general result.

Result: The quadratic form: $ax_1^2 + 2bx_1x_2 + cx_2^2$, is positive (respectively negative) definite subject to the linear constraint: $Ux_1 + Vx_2 = 0$, if and only if the determinant given below is negative (respectively positive).

$$det \begin{bmatrix} 0 & U & V \\ U & a & b \\ V & b & c \end{bmatrix}, \qquad (10.11)$$

where the middle matrix in the quadratic form from Eq. (10.8) is present in the lower right corner of the matrix in Eq. (10.11). It is 'bordered' by the coefficients (U, V) of the constraints with a zero at the top left corner.

This result readily generalizes. A quadratic form in n variables $x = (x_1, x_2, \ldots, x_n)$ defined as $x'Ax$ is positive (respectively negative) definite subject to the linear constraint $W'x = 0$, (where $W = (w_1, w_2, \ldots, w_n)$ is a column vector) if and only if the determinant of a bordered matrix $det(H)$ given below is negative (respectively positive). We define:

$$H = \begin{bmatrix} 0 & W' \\ W & A \end{bmatrix}. \tag{10.12}$$

10.3 Bilinear Form

In the above discussion of quadratic forms, if we define y as an $n \times 1$ vector, $x'Ay$ is called a bilinear form. This is also a scalar. Let e_k denote a column vector of all zeros except for the k-th element, which is unity. The individual elements a_{ij} of the matrix A are obviously scalar. Now verify that $a_{ij} = e_i'Ae_j$, can be worked out in detail by the rules of matrix multiplication and seen to be a bilinear form.

10.4 Similar Matrices

The $n \times n$ matrix A is 'similar' to another $n \times n$ matrix C if there exists a nonsingular matrix B such that

$$A = B^{-1}CB, \tag{10.13}$$

holds true even when we admit complex number entries in both matrices. One can think of A as a cheese sandwich C between two breads B^{-1} and B.

We have alluded to the property of 'similar' matrices that they have identical eigenvalues. Actually, similar matrices also have identical characteristic equation and identical trace.

Regarding eigenvectors of similar matrices matrix theory states the following result. If z is an eigenvector of A with eigenvalue λ and Eq. (10.13) holds, then

$$(B\,z) \text{ is an eigenvector of C with the eigenvalue } \lambda. \tag{10.14}$$

That is, the bread matrix B involved in the definition of the matrix C 'similar' to original A allows us to write the eigenvector of C as $(B\,z)$.

10.4.1 *Diagonalizable Matrix*

One of the uses of the concept of 'similar' matrices comes from the following property. A matrix A is said to be 'diagonalizable' if it is 'similar' to a diagonal matrix.

If A is a nonsingular symmetric matrix with distinct eigenvalues, we know from Eq. (9.19) that it can be decomposed as $A = G\Lambda G'$. We have also seen that G is an orthogonal matrix with the property $G^{-1} = G'$. Hence it is easy to see from Eq. (10.13) that such A is 'similar' to a diagonal matrix Λ implying that A is diagonalizable.

Every square matrix is not symmetric with distinct eigenvalues. In general, the matrix theory says that a square matrix is diagonalizable, if and only if it has linearly independent eigenvectors. That is, if the rank of the matrix of eigenvectors, $rk(G) = n$, is full, then A is diagonalizable.

10.5   Identity Matrix and Canonical Basis

The Euclidean vector space $\Re$ of dimension n can be readily constructed from $n \times 1$ unit vectors $e_1 = (1, 0, \ldots, 0)'$, $e_2 = (0, 1, 0, \ldots, 0)'$, and $e_i = (0, 0, \ldots, 1, \ldots, 0)'$, with 1 in the i-th position, and $e_n = (0, \ldots, 1)'$. The unit vectors $e_i, i = 1, \ldots n$ are called the canonical basis of the Euclidean vector space.

The identity matrix $I_n = \{\delta_{i,j}\}$, where $\delta_{i,j} = 1$ when $i = j$ and $\delta_{i,j} = 0$, when $i \neq j$. The delta function defining the identity matrix is called the Kronecker delta. Note that I_n is formed from the the vectors $e_i, i = 1, \ldots, n$.

The identity matrix plays the role of number 1 in scalar algebra. For example, the simple multiplication 5*3=15 can be rewritten as 5*1*3=15. That is, insertion of the number 1 does not change the result. Often insertion of a suitable identity matrix I between matrix multiplications (discussed in Sec. 4.4) also does not change the result.

Note that e_i can be thought of as the eigenvectors of I_n associated with the eigenvalue 1. The characteristic polynomial here is $(\lambda - 1)^n$, known as the 'monic polynomial' of degree n, and where the same root $(\lambda = 1)$ is repeated n times.

What about the canonical basis for the usual square matrices A? It is clear that eigenvectors can help here. However we need to generalize them as follows.

10.6   Generalized Eigenvectors and Chains

Since certain eigenvalues can repeat themselves, we need a way of seamlessly handling repeated eigenvalues of multiplicity m. Mathematicians have devised concepts called 'generalized eigenvectors' and 'chains' for this purpose.

A column vector z_m is a generalized eigenvector of rank m associated with the eigenvalue λ of the n square matrix A if

$$(A - \lambda I_n)^m z_m = 0, \text{ but } (A - \lambda I_n)^{m-1} z_m \neq 0. \tag{10.15}$$

A 'chain' associated with the eigenvalue λ generated by a generalized eigenvector of rank m is a set of eigenvectors $(z_m, z_{m-1}, \ldots z_1)$. It is defined recursively (backwards) as

$$z_j = (A - \lambda I_n)^m z_{j+1}, \text{ where } (j = m - 1, m - 2, \ldots, 1). \tag{10.16}$$

A chain of length one occurs when all eigenvalues are distinct. In that case, the eigenvectors provide a meaningful basis for the column space.

An algorithm for finding the **canonical basis** when some eigenvalues of A repeat is somewhat tedious. If an eigenvalue repeats m times we need to determine the integer p such that the rank $\text{rk}(A - \lambda I_n)^p = n - m$. Fortunately, the algorithm need not be learned these days, since R finds the correct matrix of eigenvectors G with some repeated columns when some eigenvalues repeat themselves. Hence the canonical basis can be simply obtained by simply ignoring the repeated columns produced by R.

The **minimal polynomial** associated with the $n \times n$ matrix A is defined as the monic polynomial of least degree satisfying $m_m(\lambda) = 0$. Let A have the q distinct eigenvalues: $\lambda_j, (j = 1, \ldots, q)$, with respective repetition times as $p_j, (j = 1, \ldots, q)$, where $\Sigma_{j=1}^q p_j = n$. Now the minimal polynomial is given by:

$$m_m(\lambda) = (\lambda - \lambda_1)^{p_1} (\lambda - \lambda_2)^{p_2} \ldots (\lambda - \lambda_q)^{p_q}. \tag{10.17}$$

In elementary algebra, one learns how to divide one polynomial by another polynomial by brute force. That is, the division keeps the unknown λ intact, without having to know its numerical value. Accordingly, let us consider the characteristic polynomial $f(\lambda) = 0$. If we replace λ by the original matrix A, the right side of the equation is obviously a null matrix of order $n \times n$ upon such replacement. Now matrix theory states that the minimal polynomial $m_m(\lambda)$ divides such characteristic polynomial $f(\lambda) = 0$.

10.7   Jordan Canonical Form

When eigenvalues repeat themselves things become a bit tedious to handle, because the matrix is not diagonalizable. However, but there is nothing conceptually difficult about this situation. Matrix theory defines certain 'blocks' of matrices to deal with repeated eigenvalues. A Jordan block is defined as a bidiagonal matrix with all elements zero, except for those along the main diagonal ($=\lambda$) and for the diagonal above the main diagonal ($=1$) to the right. This matrix has only one linearly independent eigenvector. For example, the $h \times h$ Jordan block matrix is defined as:

$$
J_h(\lambda) = \begin{bmatrix} \lambda & 1 & 0 & \ldots & 0 & 0 \\ 0 & \lambda & 1 & \ldots & 0 & 0 \\ \vdots & \vdots & \vdots & \vdots & \vdots & \vdots \\ 0 & 0 & 0 & \ldots & \lambda & 1 \\ 0 & 0 & 0 & \ldots & 0 & \lambda \end{bmatrix}. \tag{10.18}
$$

A matrix is said to be in Jordan canonical form if it can be partitioned into a block diagonal matrix with matrices $J_{h_1}(\lambda_1), J_{h_2}(\lambda_2), \ldots, J_{h_m}(\lambda_m)$ along its diagonal, where $\Sigma_{i=1}^m h_i = n$. The notation clarifies that we can have m distinct eigenvalues λ_i allowing us to consider repeats of distinct eigenvalues.

The key result of matrix theory called 'Jordan decomposition theorem' is that every $n \times n$ matrix A is 'similar' to an $m \times m$ block diagonal matrix $J = diag[J_j(\lambda_j)]$ in Jordan canonical form:

$$
A = PJP^{-1}, \tag{10.19}
$$

where P is a so-called 'modal matrix' which arranges the repeated eigenvalues according to following rules. The modal matrix P has the same order $n \times n$ as A containing all canonical basis vectors along its columns. If there are chains of longer length than one, we place the chains of length one before them. We line them up from the smallest to the largest length from left to right. Of course, such Jordan Form is not unique to the extent that the ordering is optional. However, we shall see that practitioners can simply accept the ordering provided by R. The theorem states that by using similarity transformations every square matrix can be transformed into the almost diagonal Jordan canonical form. A motivation based on a 4×4 matrix is available at

http://en.wikipedia.org/wiki/Jordan_normal_form

Section 3.33.2 of Marcus and Minc (1964) mentions some tricks for making off diagonal elements of the Jordan bi-diagonal matrix negligibly small by choosing certain diagonal matrix defined as:

$$E = diag(1, \delta, \ldots, \delta^{n-1}), \quad \delta > 0. \tag{10.20}$$

If J is a Jordan bidiagonal form, one can replace P used in Eq. (10.19) by (PE), which multiplies each term above the main diagonal by δ. Clearly, the trick amounts to choosing a small δ making the ones above the main diagonal of a typical Jordan form negligibly small. Of course, we cannot choose too small δ since we must keep $\delta^{n-1} \neq 0$ in numerical terms in the process of matrix multiplications in a computer. In purely algebraic terms δ^{n-1} remains nonzero even for large n.

Several examples of Jordan canonical matrices are on the Internet available at

`http://www.ms.uky.edu/~lee/amspekulin/jordan_canonical_form.pdf`
and at
`http://www.mast.queensu.ca/~math211/m211oh/m211oh89.pdf`

Weintraub (2008) and Section 10 of Chapter XIV Gantmacher (1959) discusses applications of Jordan Canonical form to finding solutions of differential equations. A collection of functions to support matrix differential calculus as presented in Magnus and Neudecker (1999) are available as R functions in the package 'matrixcalc', Novomestky (2008).

Recall the property [8] from Sec. 6.2 that the determinant of an upper or lower triangular matrix is the product of its diagonal elements. This allows us to construct a square matrix with known determinant $A - \lambda I_n$ of the characteristic equation based on Eq. (9.2). Consider an upper triangular matrix as an R object named 'A' in the following snippet.

```
# R program snippet 10.7.1 is next.
A=matrix(c(2,0,0,0, 0,2,0,0, 1,10,2,0, -3,4,0,3),4)
A# view A
cumprod(diag(A)) #view product of diagonals
det(A) #view determinant
```

Now we verify the property [8] for this example in the following output from R.

```
> A# view A
     [,1] [,2] [,3] [,4]
```

```
[1,]    2   0    1   -3
[2,]    0   2   10    4
[3,]    0   0    2    0
[4,]    0   0    0    3
> cumprod(diag(A)) #view product of diagonals
[1]  2  4  8 24
> det(A) #view determinant
[1] 24
```

Clearly the characteristic equation for this example is

$$(2 - \lambda)(2 - \lambda)(2 - \lambda)(3 - \lambda) = (\lambda - 2)^3(\lambda - 3) \tag{10.21}$$

$$= \lambda^4 - 9\lambda^3 + 30\lambda^2 - 44\lambda + 24 \tag{10.22}$$

with four roots having only two distinct eigenvalues 2 and 3. Now let us use R to compute the eigenvalues and eigenvector of this 4×4 matrix in the snippet 10.7.2 which needs the snippet 10.7.1 in the memory of R.

```
# R program snippet 10.7.2 is next.
#R.snippet get the previous snippet in memory of R
ea=eigen(A); ea$val #view eigenvalues
G=ea$vec;G #view matrix of eigenvectors
library(fBasics)
rk(G) #rank of G
```

Note that the matrix of eigenvectors is not of full rank 4, but is deficient. In other words the eigenvectors are linearly dependent, implying that this 4×4 matrix is NOT diagonalizable.

```
> ea=eigen(A); ea$val #view eigenvalues
[1] 3 2 2 2
> G=ea$vec;round(G,4) #view matrix of eigenvectors
         [,1] [,2] [,3]     [,4]
[1,] -0.5883    1    0 -0.0995
[2,]  0.7845    0    1 -0.9950
[3,]  0.0000    0    0  0.0000
[4,]  0.1961    0    0  0.0000
> rk(G) #rank of G
[1] 3
```

Visual inspection of G shows that the third row has all zeros, implying that its determinant is zero and the matrix is singular.

Prof. Carl Lee of the University of Kentucky notes at his website

`http://www.ms.uky.edu/~lee/amspekulin/jordan_canonical_form.pdf`

that all possible 4×4 bi-diagonal Jordan matrices based on Eq. (10.18) where the four diagonal terms are possibly distinct (but possibly not) and the three supra-diagonal terms (above the main diagonal) can have following numbers: (0,0,0), (1,0,0), (1,1,0), (1,0,1) and (1,1,1).

Prof. Lee also notes that the 4×4 'A' matrix of our snippet is 'almost diagonalizable,' since it is similar to an almost diagonal Jordan matrix J defined as:

$$J = \begin{bmatrix} 2 & 0 & 0 & 0 \\ 0 & 2 & 1 & 0 \\ 0 & 0 & 2 & 0 \\ 0 & 0 & 0 & 3 \end{bmatrix}. \tag{10.23}$$

We can verify the 'almost diagonalizable' claim in R as follows. First we define Prof. Lee's matrix 'P' in R. Construction of P from the matrix theory of Jordan canonical forms seems difficult since the eigenvectors found by R do not coincide with those found by Prof. Lee, forcing us to use his reported P matrix. Next, we invert it, and compute $P^{-1}AP$.

```
# R program snippet 10.7.3 is next.
P=matrix(c(0,1,0,0,  1,10,0,0,  0,0,1,0,
-3,4,0,1),4); P #view P matrix
Pinv=solve(P); Pinv# view P inverse
J=Pinv %*% A %*% P #find the similar matrix
J #view J
```

Now we see the P matrix, its inverse 'Pinv' and the matrix multiplication from Eq. (10.13) understood by R.

```
> P=matrix(c(0,1,0,0,  1,10,0,0,  0,0,1,0,
+ -3,4,0,1),4); P #view P matrix
     [,1] [,2] [,3] [,4]
[1,]   0    1    0   -3
[2,]   1   10    0    4
[3,]   0    0    1    0
[4,]   0    0    0    1
> Pinv=solve(P); Pinv# view P inverse
     [,1] [,2] [,3] [,4]
```

```
[1,]  -10   1    0   -34
[2,]    1   0    0    3
[3,]    0   0    1    0
[4,]    0   0    0    1
> J=Pinv %*% A %*% P #find the similar matrix
> J #view J
      [,1] [,2] [,3] [,4]
[1,]    2    0    0    0
[2,]    0    2    1    0
[3,]    0    0    2    0
[4,]    0    0    0    3
```

The last matrix in the R output above shows that it is indeed exactly the same as the J in Eq. (10.23).

Our next task is to use R to check whether we can make the supra diagonal term 1 along the second row 'small' (=0.01, say) by using the E matrix from Eq. (10.20).

```
# R program snippet 10.7.4 is next.
del=0.01;del #view delta
E=diag(c(1,del,del^2,del^3)); E#view E
PE=P%*%E #view PE matrix multiplication
PEinv=solve(PE) #inverse of PE
J2=PEinv %*% A %*% PE #compute new bidiagonal
round(J2,5) #note that supra diagonal 1 is made small
```

The idea of course is to use (PE) instead of P to force the only off diagonal value 1 to be equal to $\delta = 0.01$.

```
> del=0.01;del #view delta
[1] 0.01
> E=diag(c(1,del,del^2,del^3));E#view E
      [,1] [,2]  [,3]  [,4]
[1,]    1 0.00 0e+00 0e+00
[2,]    0 0.01 0e+00 0e+00
[3,]    0 0.00 1e-04 0e+00
[4,]    0 0.00 0e+00 1e-06
> PE=P%*%E #view PE matrix multiplication
> PEinv=solve(PE) #inverse of PE
> J2=PEinv %*% A %*% PE #compute new bidiagonal
```

```
> round(J2,5) #note that supra diagonal 1 is made small
     [,1] [,2] [,3] [,4]
[1,]    2    0 0.00    0
[2,]    0    2 0.01    0
[3,]    0    0 2.00    0
[4,]    0    0 0.00    3
```

Note that the original matrix A is now similar to an almost diagonal matrix, since we have successfully reduced the only off diagonal value 1 (along the second row and third column) to be equal to $\delta = 0.01$, instead of 1. Thus we are able to almost diagonalize a matrix even when eigenvalues repeat themselves by using the Jordan decomposition theorem and related ideas.

Chapter 11

Hermitian, Normal and Positive Definite Matrices

11.1 Inner Product Admitting Complex Numbers

Section 3.1.1 discusses the Euclidean inner product, whereas Sec. 4.2 discusses matrices with complex numbers. Now we use the notion of quadratic and bilinear forms to define a more general notion of inner product of vectors, where the vector elements can be complex numbers.

If x, y are $n \times 1$ column vectors and $x'y = x'I_ny$ always holds, where I_n is the matrix in the middle.

Denote the vector of complex conjugates of y as $\bar{y}$. The Euclidean inner (dot) product

$$< x \bullet y >= x'I_n\bar{y} = \bar{y}'x \qquad (11.1)$$

Now we show how to use complex numbers in R by using number 1 and i or '1i' for the imaginary symbol i. The use of this symbol creates R objects of the class 'complex'. In the following snippet 11.1.1 we create 'y' and 'ybar' as two complex vectors and compute their inner product.

```
# R program snippet 11.1.1 is next.
y=c(1,2,3+4*1i,5);y #define y
ybar=c(1,2,3-4*1i,5);ybar #define ybar
t(y)%*%ybar #inner product
t(ybar)%*%y#inner product second way
1^2+2^2+3^2+4^2+5^2
```

Note that '4*1i' with a (*) is needed to properly define a complex number. '41i' will not do. The inner product is $1^2 + 2^2 + 3^2 + 4^2 + 5^2 = 55$.

```
> y=c(1,2,3+4*1i,5);y #define y
[1] 1+0i 2+0i 3+4i 5+0i
```

```
> ybar=c(1,2,3-4*1i,5);ybar #define ybar
[1] 1+0i 2+0i 3-4i 5+0i
> t(y)%*%ybar #inner product
       [,1]
[1,] 55+0i
> t(ybar)%*%y#inner product second way
       [,1]
[1,] 55+0i
> 1^2+2^2+3^2+4^2+5^2
[1] 55
```

In general we can have an $n \times n$ nonsingular matrix D of weights. Then a general inner product is defined as:

$$< x \bullet y >_D = Dx'\bar{D}\bar{y} = \bar{D}\bar{y}'Dx, \qquad (11.2)$$

where conjugate matrices are denoted by $\bar{D}$. Note that the Euclidean inner product is a special case when $D = I_n$,

The inner product of x with itself is real and positive as long as it is not a null vector. However $< x \bullet x >_D = 0$ if and only if $x = 0$. If c is a possibly complex scalar, $< cx \bullet y >_D = c < x \bullet y >_D$. However, if c is attached to y then the answer involves $\bar{c}$. That is, $< x \bullet cy >_D = \bar{c} < x \bullet y >_D$.

The Cauchy-Schwartz inequality of Sec. 7.1.1 when complex numbers are admitted and more general inner products are considered becomes:

$$| < x \bullet y >_D |^2 \leq | < x \bullet x >_D | | < y \bullet y >_D |. \qquad (11.3)$$

where the absolute values are denoted by $|.|$. In R the absolute values of complex numbers are computed by the R function 'abs.'

Now we use R to check the inequality upon recalling two complex number objects 'y' in place of x, and 'ybar' in place of y in Eq. (11.3) defined in the previous snippet 11.1.1.

```
# R program snippet 11.1.2 (inequality check) is next.
LHS=(t(y)%*%ybar)^2;LHS  #left side
R1=abs(t(y)%*%y);R1 #first term right side
R2=abs(t(ybar)%*%ybar)#second term on right side
RHS=R1*R2; RHS
```

The following output based on R program snippet 11.1.2 shows that left side of Eq. (11.3) equals $55^2 = 3025$ which exceeds the product of two absolute values (=1105).

```
> LHS=(t(y)%*%ybar)^2;LHS  #left side
       [,1]
[1,] 3025+0i
> R1=abs(t(y)%*%y);R1 #first term right side
       [,1]
[1,] 33.24154
> R2=abs(t(ybar)%*%ybar)#second term on right side
> RHS=R1*R2; RHS
       [,1]
[1,] 1105
> 55^2
[1] 3025
```

11.2 Normal and Hermitian Matrices

The symmetric matrices containing real numbers are of interest in many applications. For example the correlation matrix is real symmetric satisfying $A = A'$. When complex numbers are admitted, practitioners like to replace the adjective 'real symmetric' with 'Hermitian.' This is because they simplify exposition.

Definition: Matrix $A = \{a_{ij}\}$ is Hermitian if and only if $a_{ij} = \bar{a}_{ij}$ for all pairs of values of i, j. When $i = j$ we have diagonal entries of the Hermitian matrix and $a_{ii} = \bar{a}_{ii}$ implies that all diagonals are real numbers. Also, if A is real symmetric matrix, then $a_{ij} = \bar{a}_{ij}$ means that off-diagonal entries are also all real numbers. This is why we can replace the adjective 'real symmetric' with 'Hermitian.'

Section 4.2, mentions A^* as the complex conjugate transpose of A. If conjugate matrices are denoted by $\bar{A}$, then we have the definition of a Hermitian matrix as $A^H = \bar{A}'$. Now define a 2×2 matrix A containing a complex number and then define $\bar{A}$ as its complex conjugate, obtained by replacing the the complex number with its complex conjugate number.

```
# R program snippet 11.2.1 is next.
A=matrix(c(1,2,3+4*1i,6),2); A#view A
Abar=matrix(c(1,2,3-4*1i,6),2); Abar#view Abar

> A=matrix(c(1,2,3+4*1i,6),2); A#view A
       [,1] [,2]
[1,] 1+0i 3+4i
```

```
[2,] 2+0i 6+0i
> Abar=matrix(c(1,2,3-4*1i,6),2); Abar#view Abar
     [,1] [,2]
[1,] 1+0i 3-4i
[2,] 2+0i 6+0i
```

Chapter 10 defined the concept of 'similar' matrices, with a focus on those that are 'similar' to diagonal matrices. The so-called normal matrices are not only 'similar' to diagonal matrices, but also possess a canonical basis of eigenvectors.

Consider following definitions using A^H where the matrix elements can be complex numbers:

a) Normal: Matrix A is normal if

$$AA^H = A^H A. \tag{11.4}$$

Consider the A and $\bar{A}$ of the previous snippet 11.2.1 and define R object 'Astar' as the transpose of $\bar{A}$. We also compute the left and right hand sides of Eq. (11.4) called LHS and RHS, respectively and note that they are not equal to each other.

R program snippet **11.2.2** (faulty normal) is next.
```
Astar=t(Abar); Astar #view A*
LHS=A%*%Astar;LHS #left side matrix
RHS=Astar%*%A; RHS #right side matrix
```

We now check if the two sides of equation are not equal.

```
> Astar=t(Abar); Astar #view A*
     [,1] [,2]
[1,] 1+0i 2+0i
[2,] 3-4i 6+0i
> LHS=A%*%Astar;LHS #left side matrix
       [,1]    [,2]
[1,] 26+ 0i 20+24i
[2,] 20-24i 40+ 0i
> RHS=Astar%*%A; RHS #right side matrix
      [,1]   [,2]
[1,]  5+0i 15+4i
[2,] 15-4i 61+0i
```

We conclude that A is not normal.

Next, Let us consider another example of matrix B which happens to be symmetric. It has a single complex number along the diagonal. There might be something wrong with B, since the imaginary number occurs alone without its complex conjugate: (1-4 i).

R program snippet **11.2.3** (absent conjugate) is next.
```
B=matrix(c(1,2,2,1+4*1i),2); B#view B
Bbar=matrix(c(1,2,2,1+4*1i),2); Bbar#view Bbar
Bstar=t(Bbar)
LHS=B%*%Bstar;LHS #left side matrix
RHS=Bstar%*%B; RHS #right side matrix
```

R reports the following left and right sides.

```
> LHS=B%*%Bstar;LHS #left side matrix
     [,1]   [,2]
[1,] 5+0i   4+8i
[2,] 4+8i -11+8i
> RHS=Bstar%*%B; RHS #right side matrix
     [,1]   [,2]
[1,] 5+0i   4+8i
[2,] 4+8i -11+8i
```

We are tempted to conclude that the B matrix is normal, since it satisfies the definition in Eq. (11.4). However, remember that it may be a defective complex object (in the sense that a complex number appears without its conjugate being also present).

b) Hermitian: Matrix A is Hermitian if

$$A^H = A. \tag{11.5}$$

Now apply the test of Eq. (11.5) to the matrix 'B' in the previous R snippet 11.2.3. The R output shows that the two sides are indeed equal.

```
> B
     [,1] [,2]
[1,] 1+0i 2+0i
[2,] 2+0i 1+4i
> Bstar
     [,1] [,2]
[1,] 1+0i 2+0i
[2,] 2+0i 1+4i
```

We conclude that the matrix R object 'B' is Hermitian, except that its definition might be defective.

The Hermitian matrices are interesting because matrix theory claims that all its eigenvalues are real numbers even if the original matrix contains imaginary numbers. This does not seem to hold for our example. The R function 'eigen' reports that the two eigenvalues of B are both (1+2 i), which are not real numbers. This happens because our matrix object 'B' is defective with a singleton complex number along its diagonal (without its conjugate).

Consider an example where the complex numbers appear in off diagonal locations, as in the R object 'C' below. Note that C is a well defined complex object since the complex number does appear with its complex conjugate on the other side of the diagonal of C.

```
# R program snippet 11.2.4 (conjugate present) is next.
C=matrix(c(1,2+2*1i,2-2*1i,4),2); C#view C
Cbar=matrix(c(1,2-2*1i,2+2*1i,4),2); Cbar#view Cbar
Cstar=t(Cbar)
LHS=C%*%Cstar;LHS #left side matrix
RHS=Cstar%*%C; RHS #right side matrix
eigen(C)$values  #report eigenvalues
```

R reports the following left and right sides.

```
> C=matrix(c(1,2+2*1i,2-2*1i,4),2); C#view C
     [,1] [,2]
[1,] 1+0i 2-2i
[2,] 2+2i 4+0i
> Cbar=matrix(c(1,2-2*1i,2+2*1i,4),2); Cbar#view Cbar
     [,1] [,2]
[1,] 1+0i 2+2i
[2,] 2-2i 4+0i
> Cstar=t(Cbar)
> LHS=C%*%Cstar;LHS #left side matrix
        [,1]    [,2]
[1,]   9+ 0i 10-10i
[2,] 10+10i 24+ 0i
> RHS=Cstar%*%C; RHS #right side matrix
        [,1]    [,2]
[1,]   9+ 0i 10-10i
```

```
[2,]  10+10i  24+  0i
> eigen(C)$values
[1]   5.7015621  -0.7015621
```

Since the R matrix object 'C' satisfies Eq. (11.5) we conclude that it is Hermitian.

If A is a Hermitian matrix containing all real numbers, A is a symmetric matrix.

Hermitian matrix is also normal and several results for normal matrices are available.

A matrix A is Hermitian iff we can write an inner product $< Ax \bullet x >$ as a real number for all $n \times 1$ vectors x.

c) Unitary: Matrix A is unitary if its inverse equals its Hermitian transpose.

$$A^H = A^{-1}, \tag{11.6}$$

where A^H is the conjugate transpose of A. Unitary matrices are orthogonal on the complex field $\mathscr{C}$ and its eigenvalues are on the unit circle. A unitary matrix having all real valued elements is called orthogonal matrix (whose inverse equals its transpose).

Now we check that the R matrix object 'C' does satisfy Eq. (11.6) from the following R output.

```
> C
      [,1]  [,2]
[1,]  1+0i  2-2i
[2,]  2+2i  4+0i
> Cstar
      [,1]  [,2]
[1,]  1+0i  2-2i
[2,]  2+2i  4+0i
```

d) Skew-Hermitian: Matrix A is skew-hermitian if

$$A^H = -A. \tag{11.7}$$

One way to create a skew-Hermitian matrix is by writing $A + Bi$, where A is real symmetric and B is skew-symmetric.

Some matrix algebra results in this context are listed below. It would be an instructive **exercise** for the reader to verify all of them with numerical examples using R. Some snippets in Sec. 11.2 will provide hints.

(i) $(A + B)^H = A^H + B^H$.
(ii) The sum of any square matrix and its complex conjugate is a Hermitian matrix. $A + A^H = (A + A^H)^H$.
(iii) The difference of any matrix and its transposed complex conjugate is a skew-Hermitian matrix. $A - A^H = -(A - A^H)^H$.
(iv) Every square matrix can be expressed as a sum of a Hermitian and skew-Hermitian matrices. (See the previous two items).
(v) A product of a matrix with its conjugate transpose is a Hermitian matrix. $AA^H = (AA^H)^H$.

Note that Hermitian, Unitary or Skew-Hermitian matrices are also normal matrices. Alternative definition of normal matrix using unitary matrices is that: A matrix A is normal iff there exists a unitary matrix B such that $BAB^H = D$ is a diagonal matrix.

Matrix theory contains an interesting result regarding individual elements $A = (a_{ij})$ and eigenvalues λ of a normal matrix A:

$$|\lambda| \geq max_{i,j}|a_{ij}|. \tag{11.8}$$

```
# R program snippet 11.2.5 (max eigen) is next.
B=matrix(c(1,2,2,1+4*1i),2); B#view B
maxBij=max(abs(c(1,2,2,1+4*1i)))
lambda1B=abs(eigen(B)$val)[1]
lambda2B=abs(eigen(B)$val)[2]
rbind(maxBij,lambda1B,lambda2B)
C=matrix(c(1,2+2*1i,2-2*1i,4),2); C#view C
maxCij=max(abs(c(1,2+2*1i,2-2*1i,4)))
lambda1C=abs(eigen(C)$val)[1]
lambda2C=abs(eigen(C)$val)[2]
rbind(maxCij,lambda1C,lambda2C)
```

We use R to check whether Eq. (11.8) holds true for the ill-defined B and well-defined C defined in snippets 11.2.3 and 11.2.4 above. Slightly abridged output of snippet 11.2.5 follows.

```
> rbind(maxBij,lambda1B,lambda2B)
              [,1]
maxBij    4.123106
lambda1B  2.236068
lambda2B  2.236068
```

```
> rbind(maxCij,lambda1C,lambda2C)
            [,1]
maxCij    4.0000000
lambda1C 5.7015621
lambda2C 0.7015621
```

Note that none of the eigenvalues 2.236068 of B exceeds the largest absolute value (=4.123106) of all elements of B_{ij}. However, note that one of the eigenvalues 5.7 of C exceeds the largest absolute value (=4) of all elements of C_{ij}. Thus Eq. (11.8) holds true only when the complex matrix object is well defined.

Let τ_k denote the sum of all elements of a principal $k \times k$ submatrix of A. Matrix theory proves that:

$$|\lambda| \geq max(|\tau_k|/k), \tag{11.9}$$

holds for normal matrices. Marcus and Minc (1964) has several advanced results regarding normal matrices in their Section 3.5. See also Pettofrezzo (1978) for matrix transformations and Hermitian matrices.

11.3 Real Symmetric and Positive Definite Matrices

Non-symmetric square matrices containing only real numbers as entries $A \in \Re$, can still have imaginary eigenvalues, because the roots of polynomials with real coefficients can be imaginary numbers. In the following snippet 11.3.1, we define 'A1' as a 3×3 matrix from a set of randomly chosen 9 positive integers.

R program snippet **11.3.1** (A1 evalues) is next.
```
set.seed(531); a=sample(4:620)
A1=matrix(a[1:9],3);A1 #view A1
eigen(A1)$valu #view eigenvalues
```

The output shows that the last two eigenvalues are a pair of complex conjugates.

```
> A1=matrix(a[1:9],3);A1 #view A1
     [,1] [,2] [,3]
[1,]    4  297  371
[2,]  345  110  398
[3,]  193  506  198
```

```
> eigen(A1)$valu #view eigenvalues
[1]   825.2570+ 0.0000i -256.6285+72.5726i -256.6285-72.5726i
```

Now we construct a 2×2 matrix 'A2', call an R package called 'matrixcalc,' Novomestky (2008), (assuming it is already downloaded from the Internet) by using the 'library' command and then use the function 'is.positive.definite' of that package to test whether A1 and A2 are positive definite.

```
# R program snippet 11.3.2 (positive definite?) is next.
library(matrixcalc)
set.seed(531); a=sample(4:620)
A2=matrix(a[1:4],2); A2 #view A2
eigen(A2)$valu #view eigenvalues
library(matrixcalc)
is.positive.definite(A1)
is.positive.definite(A2)
```

Since A1 defined in 11.3.1 has imaginary eigenvalues, it does not make sense to compute whether all eigenvalues are positive. Accordingly the function 'is.positive.definite' of the R package correctly refuses to answer. By contrast the matrix A2 has a negative eigenvalue ($= -146.2276$) and the package correctly answers 'FALSE' implying that it is not positive definite.

```
> A2=matrix(a[1:4],2); A2 #view A2
      [,1] [,2]
[1,]    4  193
[2,]  345  297
> eigen(A2)$valu #view eigenvalues
[1]   447.2276 -146.2276
> library(matrixcalc)
> is.positive.definite(A1)
Error in eval > tol : invalid comparison with complex values
> is.positive.definite(A2)
[1] FALSE
```

Note from the output of snippet 11.3.2 that A2 is not symmetric. We can consider its symmetric part by using the formula $(1/2)(A + A')$ mentioned in Sec. 4.5.4. In the following snippet 11.3.3 we use 't' to compute the transpose of A2.

\# R program snippet **11.3.3** (make symmetric) is next.

```
#R.snippet, previous snippet in memory?
A3=0.5*(A2+t(A2)); A3#view A3
is.symmetric.matrix(A3)
eigen(A3)$valu #view eigenvalues
is.positive.definite(A3)
```

```
> A3=0.5*(A2+t(A2)); A3#view A3
     [,1] [,2]
[1,]    4  269
[2,]  269  297
> is.symmetric.matrix(A3)
[1] TRUE
> eigen(A3)$valu #view eigenvalues
[1]   456.8058 -155.8058
> is.positive.definite(A3)
[1] FALSE
> solve(A3)
                [,1]            [,2]
[1,] -0.004172931  3.779523e-03
[2,]  0.003779523 -5.620109e-05
```

Since 'is.symmetric.matrix(A3)' command to R responds with 'TRUE,' and by visual inspection A3 is a symmetric 2 × 2 matrix. Since 'solve(A3)' responds with a matrix, A3 is nonsingular. Yet A3 is not positive definite. This reminds us that a negative eigenvalue can make a matrix positive definite, and yet it can be invertible (non-singular) as long as no eigenvalue is zero.

We have seen from the existence of A3 an interesting result of matrix theory: The matrix of a quadratic form can always be chosen to be symmetric.

Real symmetric matrix eigenvalues $\lambda(A) \in \Re$ are real numbers. Hence one can determine the signs of eigenvalues. If we impose the additional requirement that A is nonsingular, with $\lambda(A) > 0$, then A has strictly positive eigenvalues, and therefore a suitably defined square root matrix $A^{1/2}$ also exists. In regression models the square root matrices of positive definite error covariance matrices are used in 'generalized least squares.' See Vinod (2008a, Chapter 10) for details.

11.3.1 Square Root of a Matrix

The square root matrix $A^{(1/2)}$ is well defined for positive definite matrices A and can be numerically obtained by using the square roots of the middle matrix in an eigenvalue-eigenvector decomposition.

```
# R program snippet 11.3.1.1 is next.
A <- matrix(c(5,1,1,3),2,2); A
ei=eigen(A) #eigenvalue vector decomposition
Asqrt=ei$vec %*% sqrt(diag(ei$va)) %*% t(ei$vec)
Asqrt #print square root matrix
Asqrt%*%Asqrt #verify that it is a square root
```

Note that snippet 11.3.1.1 needs the R function 'diag' to convert the eigenvalues into a diagonal matrix and that square root matrix replaces the diagonals by their square roots by using the R function 'sqrt.'

```
> A <- matrix(c(5,1,1,3),2,2); A
     [,1] [,2]
[1,]   5    1
[2,]   1    3
> ei=eigen(A) #eigenvalue vector decomposition
> Asqrt=ei$vec %*% sqrt(diag(ei$va)) %*% t(ei$vec)
> Asqrt #print square root matrix
           [,1]        [,2]
[1,] 2.2215792 0.2541371
[2,] 0.2541371 1.7133051
> Asqrt%*%Asqrt #verify that it is a square root
     [,1] [,2]
[1,]   5    1
[2,]   1    3
```

11.3.2 Positive Definite Hermitian Matrices

A difficulty with imaginary eigenvalues is that they lack a sign. When we admit complex numbers as entries in the matrix, $A \in \mathscr{C}$, we have to generalize the notion of real symmetric matrices. The Hermitian matrices fulfill that role, since they too have real eigenvalues, that is, $\lambda(A) \in \Re$. As with quadratic forms, we test $min[\lambda(A)] < 0, \quad \leq 0, \quad > 0$ or ≥ 0.

Given that A is a Hermitian matrix, theory contains two rules:

1) A is positive (negative) definite if all their eigenvalues are strictly positive (negative).

2) A is positive (negative) semi-definite if all their eigenvalues are non-negative (non-positive). Abbreviations 'nnd' and 'npd' are commonly used in the present context.

As an aside, note that if $A \in \mathscr{C}$, then both $B = A A^H$, and $B = A^H A$ are positive definite matrices, and the square roots $\sqrt{\lambda_i(A^H A)}$ are called singular values. These are discussed in Chap. 12.

11.3.3 *Statistical Analysis of Variance and Quadratic Forms*

Let us focus on $n \times 1$ vectors of data is measured from sample means. For example, if $n = 3, x = (2, 5, 8)$ where the average of these three numbers is $\bar{x} = 5$. Now subtracting 5 from each number, we have $y = (2 - 5, 5 - 5, 8 - 5) = (-3, 0, 3)$ the vector of deviations from the sample mean, the focus of this subsection.

Define an $n \times 1$ vector ι, all of whose elements are ones. Hence their outer product produces a $n \times n$ matrix $(\iota_n \iota'_n)$, all of whose elements are ones. Now define $J^{(1)} = (1/n)(\iota_n \iota'_n)$. Finally, define another $n \times n$ matrix

$$J = I_n - J^{(1)}, \tag{11.10}$$

designed to compute deviations from the mean in the sense that Jx should replace each element of x by its deviations from the mean: $y = Jx$. The matrix J is verified to be symmetric and idempotent in the following snippet 11.3.3.1 using R in the context of the above simple example having $n = 3$.

```
# R program snippet 11.3.3.1 J for deviations from the mean.
x=c(2,5,8);x #view x
m=length(x);m #view m
J1=matrix(1/m,m,m); J1 #view J1
J=diag(m)-J1;J #view J
JJ= J%*%J; JJ #view J squared
Jx= J%*%x; round(Jx,5) #view J times x
```

We can visually note that the J matrix is square and symmetric. We verify that it is idempotent by squaring it and visually seeing that its square called 'JJ' in the snippet equals 'J'. The snippet 11.3.3.1 also checks that matrix multiplication Jx gives the deviations from the mean $y = (-3, 0, 3)$, readily known by mental subtraction.

```
> x=c(2,5,8);x #view x
[1] 2 5 8
> m=length(x);m #view m
[1] 3
> J1=matrix(1/m,m,m); J1 #view J1
          [,1]       [,2]        [,3]
[1,] 0.3333333 0.3333333 0.3333333
[2,] 0.3333333 0.3333333 0.3333333
[3,] 0.3333333 0.3333333 0.3333333
> J=diag(m)-J1;J #view J
           [,1]        [,2]        [,3]
[1,]  0.6666667 -0.3333333 -0.3333333
[2,] -0.3333333  0.6666667 -0.3333333
[3,] -0.3333333 -0.3333333  0.6666667
> JJ= J%*%J; JJ #view J squared
           [,1]        [,2]        [,3]
[1,]  0.6666667 -0.3333333 -0.3333333
[2,] -0.3333333  0.6666667 -0.3333333
[3,] -0.3333333 -0.3333333  0.6666667
> Jx= J%*%x; round(Jx,5) #view J times x
      [,1]
[1,]    -3
[2,]     0
[3,]     3
```

Note from algebraic manipulations that $\Sigma(x - \bar{x})^2 = \Sigma x^2 - n\bar{x}^2$. Hence $\Sigma x^2 = n\bar{x}^2 + \Sigma(x - \bar{x})^2$. The sum of squares Σx^2 is defined in the snippet 11.3.3.2 as the left hand side (LHS) of this equation. We want to check if this LHS is a sum of two terms on the right hand side: RHS1 and RHS2, where RHS1 equals (3 times $\bar{y}^2$) and RHS2 equals the sum of squares of deviations from the mean).

R program snippet **11.3.3.2** is next.

```
# Bring previous snippet into memory
LHS=sum(x^2)
RHS1=3*(mean(x))^2
y=Jx
RHS2=sum(y^2)
cbind(LHS, RHS1, RHS2, RHS1+RHS2)
```

The snippet checks the result that LHS=RHS1+RHS2 for the simple example.

```
> cbind(LHS, RHS1, RHS2, RHS1+RHS2)
     LHS RHS1 RHS2
[1,]  93   75   18 93
```

The result generalizes to any sized vector x and the decomposition of sum of squares is called analysis of variance (ANOVA) in Statistics. Note that we can write the LHS as the quadratic form $x'Ix$, where the matrix in the middle is the identity matrix of dimension 3. RHS1 is a quadratic form with $z'Iz$, where z is a 3×1 vector with all elements equal to the mean of x denoted as $\bar{x}$. Now RHS2 is the quadratic form $y'Iy = x'Jx$.

The 'variance' of x is proportional to sum of squares (SS) of deviations from the mean or $y'y$. The analysis of variance (ANOVA) in Statistics seeks to partition (analyze) the sum of squares SS into k components using k distinct quadratic forms with their own distinct matrices denoted by A_1 to A_k as:

$$y'y = y'A_1y + y'A_2y + \ldots y'A_ky, \tag{11.11}$$

where the matrices satisfy:

$$I = A_1 + A_2 + \ldots + A_k. \tag{11.12}$$

where the matrices are projection matrices (symmetric and idempotent) and all are used in the construction of quadratic forms.

Define a matrix X of dimension $n \times p$ containing data on p Normally distributed random variables. We can readily define a transform Y of the X matrix which contains deviations from the mean using Eq. (11.10). Thus, we can define $Y = JX$. It is well known that Y also contains Normally distributed random variable.

Now partition the p column matrix Y into two parts $Y = [X_1|X_2]$ with p_1, p_2 columns respectively. Clearly, $p = p_1 + p_2$. Let us define a so-called hat matrix, encountered earlier in Eq. (9.23), induced by the X data matrix (inserting a subscript X) as

$$H_X = X(X'X)^{-1}X'. \tag{11.13}$$

It is easy to verify that it is symmetric and idempotent projection matrix. Clearly the partitions X_1 and X_2 will induce a similar hat matrix:

$$H_{X1} = X_1(X_1'X_1)^{-1}X_1'. \tag{11.14}$$

Now we use quadratic forms in matrices of Eq. (11.13) and Eq. (11.14) to illustrate ANOVA of Eq. (11.11). We have Eq. (11.12) with $k = 3$ written as:

$$I = A_1 + A_2 + A_3, \tag{11.15}$$
$$A_1 = [I - H_X], \quad A_2 = H_{X1}, \quad A_3 = H_X - H_{X1}.$$

This shows the importance of matrix algebra in ANOVA.

11.3.4   *Second Degree Equation and Conic Sections*

Conic sections are studied in math and geometry courses. For example, http://www.math.odu.edu/cbii/calcanim/#consec provides a beautiful animation showing three-dimensional image of the cone with the plane, as well as the corresponding two-dimensional image of the plane itself. An ellipse is a set of points in the plane, whose sum of distances from two fixed points is constant. A circle and parabola are well known conic sections. Pettofrezzo (1978) (Sec. 4.7) explains the relation between matrix algebra and conics.

$$f(x, y) = ax^2 + 2bxy + cy^2 + 2dx + 2ey + f = 0 \tag{11.16}$$
$$= axx + bxy + dx$$
$$+ byx + cyy + ey$$
$$+ dx + ey + f.$$

This is written as a quadratic form similar to Eq. (10.1)

$$w'Aw = (x, y, 1) \begin{bmatrix} a & b & d \\ b & c & e \\ d & e & f \end{bmatrix} \begin{bmatrix} x \\ y \\ 1 \end{bmatrix}, \tag{11.17}$$

where the middle matrix A is 3×3 and is known as matrix of the conic form.

Now we show how to use R to plot the conic quadratic for an example used in Pettofrezzo (1978) $f(x, y) = 5x^2 + 6xy + 5y^2 - 4x + 4y - 4 = 0$. The plot is shown in Figure 11.1 as a solid line.

The particular matrix of the conic form is:

$$\begin{bmatrix} 5 & 3 & -2 \\ 3 & 5 & 2 \\ -2 & 2 & -4 \end{bmatrix}. \tag{11.18}$$

```
# R program snippet 11.3.4.1 (conic plot) is next.
conic.plot <- function (a,b,c,d,e,f,
  xlim=c(-2.5,2.5), ylim=c(-2.5,2.5), n=30, ...) {
  x0 <- seq(xlim[1], xlim[2], length=n)
  y0 <- seq(ylim[1], ylim[2], length=n)
  x <- matrix( x0, nr=n, nc=n )
  y <- matrix( y0, nr=n, nc=n, byrow=T )
  z <- a*x^2 + b*x*y + c*y^2 + d*x + e*y + f
  contour(x0,y0,z, nlevels=1, levels=0,
  drawlabels=F, axes=TRUE,...)
}
#example
conic.plot(5,3,5,-2,2,-4, main="Conic Plots")
abline(h=0) #draws the vertical axis
abline(v=0)#draws horizontal axis
conic.plot(4,0,1,0,0,-4, add=TRUE, lt=2)
conic.plot(8,0,2,0,2.828427,-4, add=TRUE, lt=3)
```

The coordinates of a point P along original axes can be rotated through angle θ by the rotation transformation discussed in Sec. 3.4.2. After rotation, the original conic in Figure 11.1 becomes the dotted line. The Figure produced by the snippet 11.3.4.1 also shows a dashed line representing a real ellipse having equation: $4x^2 + y^2 - 4 = 0$.

The symmetric 2×2 minor in the top left corner is

$$\begin{bmatrix} 5 & 3 \\ 3 & 5 \end{bmatrix}. \tag{11.19}$$

This symmetric matrix can be placed in diagonal form by the usual eigenvalue-eigenvector decomposition of Sec. 9.5. The R commands 'A=matrix(c(5,3,3,5),2); eigen(A)' yield eigenvalues as (8, 2). If we denote K=1/sqrt(2)=0.7071068, the first eigenvector is (K, K) and second eigenvector is (-K, K). Pettofrezzo (1978) describes further results involving various conics, relating them to the matrix algebra eigenvalues.

11.4   Cholesky Decomposition

A positive definite matrix A which admits complex numbers can be uniquely factored into two as

$$A = LL^H, \tag{11.20}$$

Conic Plots

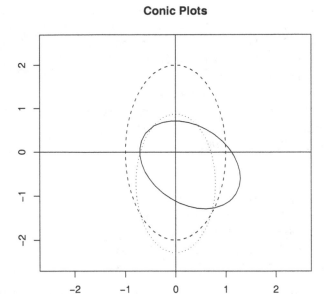

Fig. 11.1   Solid line for $5x^2 + 6xy + 5y^2 - 4x + 4y - 4 = 0$, Dashed line for $4x^2 + y^2 - 4 = 0$ and dotted line for $8x^2 + 2y^2 + 4\sqrt{2}y - 4 = 0$

where L is lower triangular with all diagonal values positive. If the elements of A are all real numbers, then an R function called 'chol' provides an upper triangular matrix as in the following snippet.

```
# R program snippet 11.4.1 is next.
A <- matrix(c(5,1,1,3),2,2); A
U <- chol(A); U
t(U) %*% U
```

The last line of the snippet 11.4.1 verifies that transpose of the upper triangular matrix 'U' produced by the R function 'chol' times 'U' will give back the original matrix A.

```
> A <- matrix(c(5,1,1,3),2,2); A
     [,1] [,2]
[1,]    5    1
```

```
[2,]    1    3
> U <- chol(A); U
        [,1]       [,2]
[1,] 2.236068 0.4472136
[2,] 0.000000 1.6733201
> t(U) %*% U
     [,1] [,2]
[1,]    5    1
[2,]    1    3
```

Anderson (2003, Sec. 7.2 and p. 631) shows that the Cholesky decomposition is useful in determining the probability distribution of covariance matrices related to the Wishart density.

11.5 Inequalities for Positive Definite Matrices

This section discusses certain inequalities involving ratios of quadratic forms with positive definite and nnd matrices are often useful in Statistics.

Let A be a real $n \times n$ non-negative definite (nnd) matrix, and let B be a real $n \times n$ positive definite matrix. Now the eigenvalues λ_i of A are defined by solving the determinantal equation: $det(A - \lambda I) = 0$, and ordering the eigenvalues from the largest to the smallest: $(\lambda_1 \geq \lambda_2 \geq \ldots \geq \lambda_n)$. Let us also consider the determinantal equation $det(A - \phi B) = 0$, with similarly ordered eigenvalues ϕ_i. Section 13.1 discusses the simultaneous diagonalization implicit here in greater detail. Now we have:

$$\lambda_n \leq [x'Ax]/[x'x] \leq \lambda_1, \tag{11.21}$$

$$\phi_n \leq [x'Ax]/[x'Bx] \leq \phi_1. \tag{11.22}$$

11.6 Hadamard Product

Instead of somewhat tedious row-column multiplication, if one multiplies two matrices $A = \{a_{ij}\}$ and $B = \{b_{ij}\}$ of identical dimensions to obtain their (simplest) multiplication we have the Hadamard product defined as:

$$C = \{c_{ij}\} = A \circ B, \quad \text{where} \quad c_{ij} = a_{ij}b_{ij}. \tag{11.23}$$

We have encountered these products in Sec. 8.9.1 in the context of estimation of heteroscedastic variances. Some properties of Hadamard products are:

(1) $A \circ B$ is a principal submatrix of the Kronecker product $A \otimes B$.
(2) Denoting the ordered eigenvalues from the largest to the smallest as $\lambda_1, \ldots \lambda_n$, assuming that A and B are nonnegative Hermitian matrices, the products of eigenvalues satisfy the inequalities:
$\lambda_n(A)\lambda_n(B) \leq \lambda_1(A \circ B) \leq \lambda_1(A)\lambda_1(B)$.
(3) The determinants satisfy the inequalities: $det(A \circ B) + det(A)det(B) \geq det(A)\Pi_{i=1}^n b_{ii} + det(B)\Pi_{i=1}^n a_{ii}$.

11.6.1 *Frobenius Product of Matrices*

Upon computing the Hadamard product $A \circ B$ of two matrices, if we simply add up all elements of $A \circ B$, we have the so-called Frobenius product. The following snippet 11.6.1.1 calls the R package for matrix calculus mentioned in Sec. 10.7 called 'matrixcalc' into memory by the 'library' command. It then uses its functions 'frobenius.prod,' and 'vec' for these computations. The 'vec' operator strings out a matrix into a long vector created by stacking its columns.

```
# R program snippet 11.6.1.1 is next.
library(matrixcalc)
A=matrix(1:4,2);A
B=matrix(11:14,2);B
AB=hadamard.prod(A,B)
sum(vec(AB))
frobenius.prod(A, B)
```

The snippet 11.6.1.1 first computes the Hadamard product by the function 'hadamard.prod' and goes on to compute the sum of all its elements by functions 'sum' and 'vec'. Note that same result 130 is also found by the R function 'frobenius.prod.'

```
> A=matrix(1:4,2);A
     [,1] [,2]
[1,]   1    3
[2,]   2    4
> B=matrix(11:14,2);B
     [,1] [,2]
[1,]   11   13
[2,]   12   14
> AB=hadamard.prod(A,B)
```

```
> sum(vec(AB))
[1] 130
> frobenius.prod(A, B)
[1] 130
```

11.7 Stochastic Matrices

A matrix $A \in M_n(\Re), A \geq 0$ is row stochastic if $AJ = J$, where J is a matrix with all elements equal to unity. The row sum of every 'row stochastic' matrix is unity. All eigenvalues of a row stochastic matrix are less than unity in absolute value.

A practical example of stochastic matrices is the transition matrix of Markov Chains.   A Markov chain satisfies the property that the state variable at time $x(t + 1)$ depends only on the state at time $x(t)$. The transition probability is defined as:

$$p_{ij} = Pr(x(t+1) = j | x(t) = i) \qquad (11.24)$$

which measures the probability p_{ij} of the state variable x being in state j at time $(t + 1)$, given that the state variable is in state i at time t. If there are n states, we can define an $n \times n$ transition matrix $P = \{p_{ij}\}$. The matrix algebra is useful here because the k-step transition probability can be computed directly as the k-th power of the transition matrix, P^k.

Matrix algebra is also useful in determining the stationary (equilibrium) probability distribution π of a Markov chain. It is a row vector and satisfies $\pi = \pi P$. It is normalized, meaning that the sum of its entries is 1, and from matrix algebra we know that π equals the *left* eigenvector of the transition matrix associated with the eigenvalue 1.

Recall from Sec. 3.4.5 that we have defined a matrix $J_n \in M_n(\Re)$, where all entries equal one. Define J_n^0 as an analogous matrix with all entries $(1/n)$. Then a matrix $A \in M_n(\Re)$ is doubly stochastic if $J_n^0 A = A J_n^0 = J_n^0$. Doubly stochastic matrices are such that their row sums and column sums equal unity. Note that permutation matrices defined in Sec. 16.4.2 are doubly stochastic.

11.8 Ratios of Quadratic Forms, Rayleigh Quotient

We have discussed maxima and minima of quadratic forms in Sec. 10.1. Statistical hypothesis testing often uses test statistics which are *ratios* of quadratic forms in Normal random variables x. Hence there is practical interest in such ratios. A related Rayleigh quotient is defined as:

$$R(A, x) = (x'Ax)/(x'x), \tag{11.25}$$

which has a quadratic form with its $n \times n$ matrix A in the numerator and the identity matrix in the denominator. The denominator in Eq. (11.25) removes the effect of scale on the ratio. That is, the Rayleigh quotient remains the same even if we multiply the x vector by any arbitrary nonzero number. This invariance is an interesting property which allows us to replace x by $u = (x'x)^{-1/2}x$ which is of unit length: $(u'u = 1)$, without loss of generality.

Let the eigenvalues of A be real numbers ordered as $\lambda_1 \geq \lambda_2 \geq \ldots \geq \lambda_n$, and let $x \neq 0$ be any arbitrary vector of real numbers. Now the global maxima and minima of the Rayleigh quotient, respectively, are given (for arbitrary x assuming that $x \neq 0$) by:

$$max[(x'Ax)/(x'x)] = \lambda_1 \quad \text{and} \quad min[(x'Ax)/(x'x)] = \lambda_n. \tag{11.26}$$

We also have the following inequalities implied by Eq. (11.26):

$$\lambda_n \leq [(x'Ax)/(x'x)] \leq \lambda_1. \tag{11.27}$$

Application: Comparison of Unemployment Rates.

Consider an application of these results. Let x denote weekly data on $n = 104$ recent unemployment rates in $k = 50$ States in USA. Let the average unemployment rate for each of the 50 States be denoted by $n \times 1$ vectors $\mu_i, (i = 1, \ldots k)$. Their average: $\mu = (1/k)\Sigma_{i=1}^k \mu_i$, is readily defined for each time period and hence is an $n \times 1$ vector.

Define the $n \times n$ outer product matrix: $O_i = (\mu_i - \mu)(\mu_i - \mu)'$, based on the deviations of each State from the national average. If O_i is zero, $\mu_i = \mu$, implying that i-th State unemployment rate equals the national average. Now define the sum of outer products: $A = \Sigma_{i=1}^k O_i$, and hence a quadratic form $(x'Ax)$. By applying Eq. (11.26) and Eq. (11.27) we can bound the interstate difference in unemployment rates by the ordered eigenvalues of A. Recall that we can normalize x to be unit vectors u and that eigenvectors are always unit vectors (by definition). Hence weights on unit eigen*vectors* associated with the largest (smallest) eigenvalues will help identify the names of those States that deviate most (least) from the national average.

However, Statisticians will ask us not to limit attention on State means only while ignoring the variability among State means due to intrinsic geographical or historical reasons. Since geography and history does not change over time, it is convenient to assume that such variabilities among States can be approximately represented by a common 'positive definite' (variance-)covariance matrix V, which could be estimated by a sum of k outer products of variances.

Note that interstate differences are more important ('significant') if they are in low-variability directions. We can down-weight high-variability directions by placing variances in the denominators of a new ratio for a more rigorous comparison between States:

$$\tilde{R}(A, x) = (x'Ax)/(x'Vx). \tag{11.28}$$

The maximum of the new ratio $\tilde{R}(A, x)$ will correctly allow for intrinsic interstate variabilities and still identify States with maximum variability of average unemployment rates from the national mean. Since V is positive definite, we can compute the largest eigenvalue $\lambda_1(V^{-1}A)$ as the upper limit (maxima) of $\tilde{R}(A, x)$. Also, we use the smallest eigenvalue $\lambda_n(V^{-1}A)$ as the lower limit (minima) of $\tilde{R}(A, x)$. These methods obviously apply to comparisons among group means of any kind.

## Chapter 12

Kronecker Products and Singular Value Decomposition

Chapter 8 discusses the matrix inverse including the case when the matrix is partitioned. Partitioning of a matrix into blocks sometimes arises naturally. For example, if we want to study and solve a bunch of similar equations at the same time, each equation can be considered a block. After putting together blocks of matrices it helps to know multiplications of blocks called Kronecker products. Chapter 6 discusses singularity of a matrix, which is related to its zero eigenvalue. Section 7.3 mentions the Singular Value Decomposition (SVD) in the context of property [8] of ranks. This chapter studies these concepts in greater detail.

12.1 Kronecker Product of Matrices

Kronecker product of matrices is a powerful notational device in multivariate analysis to study several things at the same time. We define the Kronecker product by the notation $C = A \otimes B$, where A is $T \times n$, B is $m \times p$ and C is a large matrix of dimension $Tm \times np$ with elements defined by $C = (a_{ij}B)$, where each term $a_{ij} B$ is actually a_{ij} times the whole $m \times p$ matrix.

For example, if $A = (a_{11}, a_{12}, a_{13})$ is a row vector with $T = 1$ and $n = 3$, and if

$$B = \begin{bmatrix} b_{11} & b_{12} \\ b_{21} & b_{22} \end{bmatrix} \tag{12.1}$$

we have

$$C = \left(a_{11} \begin{bmatrix} b_{11} & b_{12} \\ b_{21} & b_{22} \end{bmatrix}, \ a_{12} \begin{bmatrix} b_{11} & b_{12} \\ b_{21} & b_{22} \end{bmatrix}, \ a_{13} \begin{bmatrix} b_{11} & b_{12} \\ b_{21} & b_{22} \end{bmatrix} \right). \tag{12.2}$$

Upon multiplication of the components of A with the attached matrix B we have

$$C = \begin{bmatrix} a_{11}b_{11}, & a_{11}b_{12}, & a_{12}b_{11}, & a_{12}b_{12}, & a_{13}b_{11}, & a_{13}b_{12} \\ a_{11}b_{21}, & a_{11}b_{22}, & a_{12}b_{21}, & a_{12}b_{22}, & a_{13}b_{21}, & a_{13}b_{22} \end{bmatrix}. \tag{12.3}$$

Now we evaluate this in R. First choose A and B similar to the definitions in Eq. (12.1) and then use the R command '%x%' for a Kronecker product of A and B.

```
#R.snippet
A=matrix(1:3,1,3);A# view A
B=matrix(11:14,2);B # view B
C=matrix(21:24,2);C#view C
AkB=A%x%B; AkB#view A kronecker product B
BkC=B %x% C; BkC#view B kronecker C
```

Now the kronecker product is seen to be as suggested by Eq. (12.3)

```
> A=matrix(1:3,1,3);A# view A
     [,1] [,2] [,3]
[1,]    1    2    3
> B=matrix(11:14,2);B # view B
     [,1] [,2]
[1,]   11   13
[2,]   12   14
> C=matrix(21:24,2);C#view C
     [,1] [,2]
[1,]   21   23
[2,]   22   24
> AkB=A%x%B; AkB#view A kronecker product B
     [,1] [,2] [,3] [,4] [,5] [,6]
[1,]   11   13   22   26   33   39
[2,]   12   14   24   28   36   42
> BkC=B %x% C; BkC#view B kronecker C
     [,1] [,2] [,3] [,4]
[1,]  231  253  273  299
[2,]  242  264  286  312
[3,]  252  276  294  322
[4,]  264  288  308  336
```

In the following, let A, B, C and D be arbitrary matrices. The Kronecker product satisfies the following kinds of linearity and distributive laws.

$\left(\text{i}\right)$ $A \otimes (aB) = a(A \otimes B), a$ being a scalar;

```
#R.snippet bring in memory the A, B and C above.
a=3;a #define constant a=3
aB= a*B; aB #view scalar multiplication
AkaB=A %x% aB; AkaB; #View the left side of property (i)
aAkB=a*AkB; aAkB # view the right side
```

```
> aB= a*B; aB #view scalar multiplication
     [,1] [,2]
[1,]   33   39
[2,]   36   42
> AkaB=A %x% aB; AkaB; #View the left side of property (i)
     [,1] [,2] [,3] [,4] [,5] [,6]
[1,]   33   39   66   78   99  117
[2,]   36   42   72   84  108  126
> aAkB=a*AkB; aAkB # view the right side
     [,1] [,2] [,3] [,4] [,5] [,6]
[1,]   33   39   66   78   99  117
[2,]   36   42   72   84  108  126
```

The property clearly holds for this example.

$\left(\text{ii}\right)$ $(A+B) \otimes C = A \otimes C + B \otimes C$, A and B being of the same order.

```
#R.snippet bring A, B C into R memory
A=matrix(1:4,2);A #redefine A as 2 by 2 matrix
ApB=A+B #p denotes plus
ApBkC=ApB %x% C; ApBkC# view A+B kronicker C
AkC=A %x% C;AkC# A kronecker C
BkC=B%x%C
RHS=AkC+BkC; RHS
```

Now our hands on check of the above claim is found in the following R output.

```
> A=matrix(1:4,2);A #redefine A as 2 by 2 matrix
     [,1] [,2]
[1,]    1    3
[2,]    2    4
> ApB=A+B #p denotes plus
```

```
> ApBkC=ApB %x% C; ApBkC# view A+B kronicker C
      [,1] [,2] [,3] [,4]
[1,]   252  276  336  368
[2,]   264  288  352  384
[3,]   294  322  378  414
[4,]   308  336  396  432
> AkC=A %x% C;AkC# A kronecker C
      [,1] [,2] [,3] [,4]
[1,]    21   23   63   69
[2,]    22   24   66   72
[3,]    42   46   84   92
[4,]    44   48   88   96
> BkC=B%x%C
> RHS=AkC+BkC; RHS
      [,1] [,2] [,3] [,4]
[1,]   252  276  336  368
[2,]   264  288  352  384
[3,]   294  322  378  414
[4,]   308  336  396  432
```

$\left(\text{iii}\right)$ $A \otimes (B+C) = A \otimes B + A \otimes C$, B and C being of the same order;

```
> A %x% (B+C)
      [,1] [,2] [,3] [,4]
[1,]    32   36   96  108
[2,]    34   38  102  114
[3,]    64   72  128  144
[4,]    68   76  136  152
> A %x% B + A %x%C
      [,1] [,2] [,3] [,4]
[1,]    32   36   96  108
[2,]    34   38  102  114
[3,]    64   72  128  144
[4,]    68   76  136  152
```

$\left(\text{iv}\right)$ $A \otimes (B \otimes C) = (A \otimes B) \otimes C$;

```
> A %x% (B %x%C)
      [,1] [,2] [,3] [,4] [,5] [,6] [,7] [,8]
[1,]   231  253  273  299  693  759  819  897
```

```
[2,]   242   264   286   312   726   792   858   936
[3,]   252   276   294   322   756   828   882   966
[4,]   264   288   308   336   792   864   924  1008
[5,]   462   506   546   598   924  1012  1092  1196
[6,]   484   528   572   624   968  1056  1144  1248
[7,]   504   552   588   644  1008  1104  1176  1288
[8,]   528   576   616   672  1056  1152  1232  1344
> (A %x% B)  %x%C
       [,1]  [,2]  [,3]  [,4]  [,5]  [,6]  [,7]  [,8]
[1,]   231   253   273   299   693   759   819   897
[2,]   242   264   286   312   726   792   858   936
[3,]   252   276   294   322   756   828   882   966
[4,]   264   288   308   336   792   864   924  1008
[5,]   462   506   546   598   924  1012  1092  1196
[6,]   484   528   572   624   968  1056  1144  1248
[7,]   504   552   588   644  1008  1104  1176  1288
[8,]   528   576   616   672  1056  1152  1232  1344
```

The two large matrices are identical checking the fourth property.

$$\left(\text{V}\right) \quad (A \otimes B)' = A' \otimes B' \; ;$$ (The order does not reverse here as in simple matrix multiplication)

```
> t(A %x%B)
       [,1]  [,2]  [,3]  [,4]
[1,]    11    12    22    24
[2,]    13    14    26    28
[3,]    33    36    44    48
[4,]    39    42    52    56
> t(A) %x% t(B)
       [,1]  [,2]  [,3]  [,4]
[1,]    11    12    22    24
[2,]    13    14    26    28
[3,]    33    36    44    48
[4,]    39    42    52    56
```

We use the R function 't' to compute the transpose and check the above property hands on.

$$\left(\text{vi}\right) \quad (A \otimes B)(C \otimes D) = AC \otimes BD;$$
$$\left(\text{vii}\right) \quad (A \otimes B)^{-1} = A^{-1} \otimes B^{-1}.$$

```
> solve(A %x%B)
      [,1]  [,2]   [,3]   [,4]
[1,]    14 -13.0 -10.5   9.75
[2,]   -12  11.0   9.0  -8.25
[3,]    -7   6.5   3.5  -3.25
[4,]     6  -5.5  -3.0   2.75
> solve(A) %x% solve(B)
      [,1]  [,2]   [,3]   [,4]
[1,]    14 -13.0 -10.5   9.75
[2,]   -12  11.0   9.0  -8.25
[3,]    -7   6.5   3.5  -3.25
[4,]     6  -5.5  -3.0   2.75
```

We use the R function 'solve' to compute the inverse and check the above property hands on. Again, the order does not reverse here as in the inverse of simple matrix multiplication. Obviously we assume that A and B are square non-singular matrices.

A practical application of this result (vii) in the context of GLS estimation is worth mentioning here. Eq. (8.52) write the covariance matrix of a large system of M equations as $\tilde{\Omega} = \Sigma \otimes I$, where both Σ and I are square and nonsingular. The result (vii) allows us to write $\tilde{\Omega}^{-1} = \Sigma^{-1} \otimes I$, since the inverse of the identity is itself. Thus, instead of inverting a large $MT \times MT$ matrix $\tilde{\Omega}$ we can simply write inverse as $\Sigma^{-1} \otimes I$.

$\left(\text{viii}\right)$ $trace(A \otimes B) = (traceA)(traceB)$, A and B being square matrices;

```
#R.snippet
trace=function(A){sum(diag(A))} #new R function
trace(A %x%B)
trace(A) * trace(B)
```

The snippet starts with a one line *ad hoc* function called 'trace' to compute the trace of a square matrix by summing all of its diagonal entries. The next two lines should yield numerically identical results of the relation (viii) holds true.

```
> trace(A %x%B)
[1] 125
> trace(A) * trace(B)
[1] 125
```

We see that the trace of a Kronecker product is the simple product of the two traces.

$$\left(\text{ix}\right) \quad det(A \otimes B) = (detA)^m (detB)^n,$$

A and B being $n \times n$ and $m \times m$ matrices;

```
#R.snippet
det(A %x%B)
(det(A)^2) * (det(B)^2)
```

Note that both matrices are 2×2. Hence the determinants are raised to the power 2 before their multiplication.

```
> det(A %x%B)
[1] 16
> (det(A)^2) * (det(B)^2)
[1] 16
```

$\left(\text{x}\right)$ Rank of a Kronecker product is the product of ranks: $rank(A \otimes B) = rank(A) * rank(B)$.

```
#R.snippet
library(fBasics)
rk(A%x%B)
rk(A) * rk(B)
```

```
> rk(A%x%B)
[1] 4
> rk(A) * rk(B)
[1] 4
```

An implication of this property is that the Kronecker product is singular unless both components are nonsingular. If A is $m \times n$ and B is $p \times q$, the rank of Kronecker product

$$rank(A \otimes B) = rank(A) * rank(B) \leq min(m, n)min(p, q),$$

where the min means the smaller of the two dimension numbers. If either A or B is singular, their product is singular.

$\left(\text{xi}\right)$ Rank of a Kronecker product, $rank(A \otimes B)$, is the rank of a matrix multiplication of the Kronecker product with its transpose $rank\left[(A \otimes B)(A \otimes B)'\right]$, and it also equals the rank of the Kronecker product of the two matrices post multiplied by their own transposes, $rank(AA' \otimes BB')$.

```
#R.snippet previous snippets in memory
rk(A%x%B)
rk((A%x%B) %*% t(A%x%B))
rk((A %*%t(A)) %x% (B %*% t(B)) )
```

```
> rk(A%x%B)
[1] 4
> rk((A%x%B) %*% t(A%x%B))
[1] 4
> rk((A %*%t(A)) %x% (B %*% t(B)) )
[1] 4
```

Consider data for $T = 20$ years and for $N = 50$ States in US called a time series of cross sections (or panel data or longitudinal data). Certain data features over T time units and other features N individuals need to be studied separately. The entire data set has TN observations. It is possible to summarize into a $T \times T$ matrix A_T the over time features of the data. Similarly a $N \times N$ matrix A_N summarizes the individual (State) centered features. An interaction between the two types of effects can be viewed as a Kronecker product: $A_N \otimes A_T$. Consider a simplified model which assumes that $A_T = cI_T$, where c is a constant scalar, that is, A_T is proportional to the identity matrix. Now $A_N \otimes A_T = cA_N \otimes I_T$ allows us to estimate individual (State) features.

12.1.1   *Eigenvalues of Kronecker Products*

Eigenvalues of Kronecker Product equals the product of eigenvalues. For example, If A and B are square matrices of dimensions m and n respectively, then: $C = A \otimes B$, is an mn dimensional matrix. If the eigenvalues of A are denoted by $\lambda_i(A), i = 1, 2, \ldots, m$ and those of B are denoted by $\lambda_j(B), j = 1, 2, \ldots, n$, then the mn eigenvalues of C are simply a product of the eigenvalues $(\lambda_i(A)\lambda_j(B))$ with $i = 1, 2, \ldots, m$ and $j = 1, 2, \ldots, n$.

```
#R.snippet for eigenvalues of the Kronecker product
eigen(A %x%B)$values
eva=eigen(A)$values
evb=eigen(B)$values
c(eva*evb[1], eva*evb[2])
```

Note that we use the R function 'c' to place the eigenvalues in a combined list.

```
> eigen(A %x%B)$values
[1] 134.73544902  -9.33672089  -0.42841593   0.02968781
> eva=eigen(A)$values
> evb=eigen(B)$values
> c(eva*evb[1], eva*evb[2])
[1] 134.73544902  -9.33672089  -0.42841593   0.02968781
```

12.1.2 *Eigenvectors of Kronecker Products*

Eigenvectors of Kronecker Products contain stacked products of elements of corresponding individual eigenvectors which can be viewed as Kronecker product of matrices of eigenvectors.

Using the notation in Sec. 12.1.1, all eigenvectors of C have $k = i(j) = 1, 2, \ldots, mn$ elements each. Each element of each eigenvector of A is multiplied by the entire eigenvector of B. That is, using R it can be seen that the eigenvectors of a Kronecker product are a Kronecker product of two matrices of eigenvectors of the individual matrices.

```
#R.snippet for eigenvectors of Kronecker products.
eigen(A %x%B)$vec
evea=eigen(A)$vec
eveb=eigen(B)$vec
evea %x% eveb
```

The following R output confirms the result about eigenvectors.

```
> eigen(A %x%B)$vec
          [,1]        [,2]        [,3]        [,4]
[1,] 0.3838016   0.6168970  -0.4305939   0.6921078
[2,] 0.4156791   0.6681348   0.3669901  -0.5898752
[3,] 0.5593629  -0.2821854  -0.6275592  -0.3165889
[4,] 0.6058221  -0.3056230   0.5348613   0.2698249
> evea=eigen(A)$vec
> eveb=eigen(B)$vec
> evea %x% eveb
          [,1]        [,2]        [,3]        [,4]
[1,] 0.3838016   0.4305939   0.6168970   0.6921078
[2,] 0.4156791  -0.3669901   0.6681348  -0.5898752
```

```
[3,]  0.5593629   0.6275592  -0.2821854  -0.3165889
[4,]  0.6058221  -0.5348613  -0.3056230   0.2698249
```

Note that the last matrix is constructed from a Kronecker product of matrices of eigenvectors for A and B denoted in the snippet as 'evea' and 'eveb' respectively. The output shows that the Kronecker product of eigenvectors is that same as eigenvectors of Kronecker product except that the second and third columns are interchanged.

12.1.3   *Direct Sum of Matrices*

The Kronecker products are also called direct products. This subsection discusses analogous direct sums. If A is $n \times n$ and B is $m \times m$, their direct sum: $D = A \oplus B$ is a block diagonal $(n + m) \times (n + m)$ matrix:

$$D = A \oplus B = \begin{bmatrix} A & 0 \\ 0 & B \end{bmatrix} \tag{12.4}$$

Let us note two properties of direct sums of A_i matrices of dimension $n_i \times n_i$ before closing this section.

(1) $A \oplus B \neq B \oplus A$
(2) Direct sum is distributive. $(A_1 \oplus A_2) \oplus A_3 = A_1 \oplus (A_2 \oplus A_3)$

12.2   Singular Value Decomposition (SVD)

The SVD is an extension of eigenvalue-eigenvector decomposition given earlier in Eq. (9.19) stating that $X'X = G\Lambda G'$, where G is an orthogonal matrix of eigenvectors and Λ is a diagonal matrix of eigenvalues λ_i.

The square roots of eigenvalues $\sqrt{\lambda_i}$ are known as 'singular values'. Hence the eigenvalue-eigenvector decomposition already gives us the squares of eigenvalues. Since the diagonal matrix of eigenvalues is Λ, we use the matrix $\Lambda^{1/2}$ to denote the singular values. It is just a diagonal matrix of some numbers and R denotes it as 'd.' Section 9.5 uses R to check that the original matrix equals the matrix obtained by multiplying out the decomposed components.

Householder (1964) discusses matrix theory associated with Lanczos algorithm which converts any matrix into tri-diagonal form. The eigenvalues and singular values for tri-diagonal matrices can be computed reliably. An R package called 'irlba' by Jim Baglama and Lothar Reichel is available for

fast computation of partial SVD by implicitly-restarted Lanczos bidiago-
nalization algorithm.

An important motivation for studying SVD comes from Statistics. Let
us describe the concept of SVD in that context, without loss of generality.
Assume that X is a $T \times p$ matrix of regressors for the regression model
in matrix notation. Our snippet uses for illustration a 10×4 matrix for
brevity. In real applications T, p can be large numbers.

Consider a standard multiple regression model:

$$y = X\beta + \varepsilon, \tag{12.5}$$

where y is a $T \times 1$ vector of data on the dependent variable, β is a $p \times 1$
vector of regression coefficients, and ε is $T \times 1$ vector of unknown errors or
'disturbances' of the linear model.

The eigenvalues $\lambda_i(A)$ of the square matrix $A = X'X$ are usually distinct. Then the eigenvectors in G of A are called 'principal axes' for the
data X. These are meaningful in the sense that the first principal axis
associated with the largest eigenvalue occurs in the region of the data with
the highest 'spread' or variability or 'information.' By contrast the original basis of $\Re$ having vectors $e_i, i = 1, \ldots n$, used to measure the regressor
variables depends on the accident of choosing which variable goes first in
constructing the X matrix of regressors. That is, original data lack any
such interpretation (permitted by SVD) regarding relative data spreads.

One of the issues in regression analysis is whether the least squares
solution exists. If the X matrix is not of full column rank, then it can
be shown that the inverse $(X'X)^{-1}$ does not exist. Hence the solution
$\hat{\beta} = (X'X)^{-1}(X'y)$ cannot be computed at all.

Vinod (2008a, Sec. 1.9.1) describes how to apply SVD to X itself,
and explains how it provides a deeper understanding of the regressor data.
The R software command 'svdx=svd(X)' creates an object called svdx,
which contains three matrices representing a decomposition of X into three
matrices as:

$$X = U\Lambda^{1/2}G', \tag{12.6}$$

where U is a $T \times p$ matrix, having dimensions similar to X itself. It satisfies:
$U'U = I$. This means that U' is a 'left inverse' of U. Note, however, that
UU' does NOT equal identity. That is, U' is not a 'right inverse' of U. We
conclude that Eq. (12.6) U is not an orthogonal matrix.

The geometric interpretation of U is that it contains standardized sample principal coordinates of X. Given the multidimensional scatter of all

data, one places an ellipsoid around it. The first *principal axis* has the greatest spread as illustrated in Vinod and Ullah (1981). Each subsequent principal axes are perpendicular to the previous ones and have sequentially the greatest spread.

The matrices Λ and G in Eq. (12.6) are exactly the same as in Eq. (9.19). The middle diagonal matrix $\Lambda^{1/2}$ contains the so-called "singular values" of X, which are square roots of eigenvalues. If one of the singular values is zero, then the matrix $X'X$ is singular. Recall from Sec. 6.4 that a matrix is said to be singular if it cannot be inverted and/or its determinant is zero and/or one or more of its eigenvalues are zero. The eigenvectors in columns of G, g_i are direction cosine vectors (cosines of direction angles), which orient the i-th principal axis of X with respect to the given original axes of the X data. See the next R snippet.

```
#R snippet Verify that SVD does decompose X
set.seed(987);a=sample(10:600)
X=matrix(a[1:40],10,4);X #view data matrix
svdx=svd(X) #this object has 3 matrices u, d, and v
svdx$u %*% diag(svdx$d) %*% t(svdx$v) #the decomposition
#verify that we got the original X matrix back after the SVD
svdx$d[1]/svdx$d[ncol(X)] #condition number of X >4
```

If you give the data matrix 'X' to R, it outputs the matrix 'u', diagonals of the middle matrix 'd' and the orthogonal matrix 'v'. The user of 'svd' needs to know these names within R to correctly extract the components and remember to use the transpose of the last matrix if one wants to recreate the original matrix from the decomposition. Our notation $U, \Lambda^{1/2}, G$ appearing in Eq. (12.6) is somewhat different from the notation used by the R programmer of 'svd.'. The middle matrix is a diagonal matrix. The last line of the snippet computes the condition number of the data matrix X. Its discussion is postponed.

```
> X=matrix(a[1:40],10,4);X #view data matrix
      [,1] [,2] [,3] [,4]
 [1,]  292  117  452  263
 [2,]  594  231  297  408
 [3,]  367  108  118  191
 [4,]  349  190  436  365
 [5,]  484  123  554   75
 [6,]  146  378   15  133
```

```
   [7,]   258    86   290   587
   [8,]   213   564   291   305
   [9,]   513    66   194   477
  [10,]   423   415   110   500
> svdx=svd(X) #this object has 3 matrices u, d, and v
> svdx$u %*% diag(svdx$d) %*% t(svdx$v) #the decomposition
         [,1] [,2] [,3] [,4]
   [1,]   292   117   452   263
   [2,]   594   231   297   408
   [3,]   367   108   118   191
   [4,]   349   190   436   365
   [5,]   484   123   554    75
   [6,]   146   378    15   133
   [7,]   258    86   290   587
   [8,]   213   564   291   305
   [9,]   513    66   194   477
  [10,]   423   415   110   500
> #verify that we got the original X matrix back after the SVD
> svdx$d[1]/svdx$d[ncol(X)] #condition number of X >4
[1] 5.515602
```

Substituting SVD in Eq. (12.6), we have two alternate specifications whereby the unknown parameter vectors (to be estimated) might be denoted either by γ or by α:

$$y = X\beta + \varepsilon = U\Lambda^{1/2}G'\beta + \varepsilon = U\Lambda^{1/2}\gamma + \varepsilon = U\alpha + \varepsilon, \qquad (12.7)$$

where we have used notation γ for $G'\beta$ and $\alpha = \Lambda^{1/2}\ G'\beta$. The notation change is not a "material" change, and yet it reveals some important insights. According to the last equation, the matrix U now has the p columns for p transformed regressors. Since $U'U = I$, only the *columns* of U when they are viewed as regressors are orthogonal. It may seem that this has removed collinearity completely. It can be shown that the problem of collinearity does not go away by any change of notation. Collinearity will reveal its ugly head, when we try to estimate the β vector even from the reliably available estimates of the α vector (due to the presence of the G matrix).

12.2.1   *SVD for Complex Number Matrices*

The singular values of a matrix can obviously have imaginary numbers, since the roots of its characteristic equation (a polynomial of possibly high order) can be imaginary numbers. If the original matrix A contains complex numbers, the 'svd' function of R gives some output, but it does NOT really work. The matrix multiplication of the u d and v components does not give back the original matrix. R also offers a version of the function 'svd' called 'La.svd' which returns the same three matrices except that it replaces v by vt, the (conjugated if complex) transpose of v.

In the presence of complex numbers within a matrix, we must first define its complex conjugate $\bar{A}$ and then compute its transpose to construct a Hermitian matrix A^H as explained in Chap. 11. Next, we work with the matrix product $A^H A$ (similar to the matrix $X'X$ above) and compute its SVD by using the command 'La.svd.' This is shown in the following snippet.

```
#R.snippet
A=matrix(1:9,3);A #artificial matrix
A[2,3]=3+4*1i #make one element complex
A #view the complex valued A
s0=svd(A) #svd object called s0
s2$u %*% diag(s2$d) %*% t(s2$v) #=A?
s1=La.svd(A) #svd object called s1
#La.svd outputs the transpose of last matrix
s1$u %*% diag(s1$d) %*% s2$v #=A?
Abar=A #initialize place for conjugate A
Abar[2,3]=3-4*1i # define Abar conj matrix
AH=t(Abar); AH# define Hermitian matrix
s2=svd( AH%*%A) #svd object called s2
s2$u %*% diag(s2$d) %*% t(s2$v)#=AH?
s3=La.svd( AH%*%A) #svd object called s2
s3$u %*% diag(s3$d) %*% s3$v #=AH?
```

Note that original A matrix with a complex number in the [2,3] location does not get reproduced when we multiply out the svd decomposed outputs. Hence svd or La.svd should be used with caution when complex numbers are present.

The snippet also shows that the object $A^H A$ denoted in the snippet as AHA using the Hermitian matrix is reproduced only when one uses La.svd, not if one uses the simple svd.

```
> A=matrix(1:9,3);A #artificial matrix
      [,1] [,2] [,3]
[1,]    1    4    7
[2,]    2    5    8
[3,]    3    6    9
> A[2,3]=3+4*1i #make one element complex
> A #view the complex valued A
      [,1] [,2] [,3]
[1,] 1+0i 4+0i 7+0i
[2,] 2+0i 5+0i 3+4i
[3,] 3+0i 6+0i 9+0i
> s0=svd(A) #svd object called s0
> s2$u %*% diag(s2$d) %*% t(s2$v) #=A?
      [,1]                [,2]                [,3]
[1,] 14-0i 32.00000- 0.00000i  40.00000- 8.00000i
[2,] 32+0i 76.97797+ 0.03618i  97.24215-18.87737i
[3,] 40-8i 97.24215-18.87737i 142.71794-60.44169i
> s1=La.svd(A) #svd object called s1
> #La.svd outputs the transpose of last matrix
> s1$u %*% diag(s1$d) %*% s2$v #=A?
                   [,1]                [,2]                  [,3]
[1,] 0.8960285+0.2751959i 2.518661+0.212743i  -7.541620+1.362520i
[2,] 2.4520176-0.4123948i 3.592319-0.699409i  -4.843619-3.311038i
[3,] 2.9686318+0.1576926i 3.712258+0.382261i -10.010063+1.741839i
> Abar=A #initialize place for conjugate A
> Abar[2,3]=3-4*1i # define Abar conj matrix
> AH=t(Abar); AH# define Hermitian matrix
      [,1] [,2] [,3]
[1,] 1+0i 2+0i 3+0i
[2,] 4+0i 5+0i 6+0i
[3,] 7+0i 3-4i 9+0i
> s2=svd( AH%*%A) #svd object called s2
> s2$u %*% diag(s2$d) %*% t(s2$v)#=AH?
      [,1]                [,2]                [,3]
[1,] 14-0i 32.00000- 0.00000i  40.00000- 8.00000i
[2,] 32+0i 76.97797+ 0.03618i  97.24215-18.87737i
[3,] 40-8i 97.24215-18.87737i 142.71794-60.44169i
> s3=La.svd( AH%*%A) #svd object called s2
> s3$u %*% diag(s3$d) %*% s3$v #=AH?
```

```
     [,1]    [,2]     [,3]
[1,] 14-0i 32+ 0i  40+ 8i
[2,] 32+0i 77+ 0i  97+20i
[3,] 40-8i 97-20i 155- 0i
```

This shows that 'La.svd' applied to $A^H A$ gets back the right components so that matrix multiplication of the components gives back the original matrix.

12.3   Condition Number of a Matrix

The condition number is a concept from numerical analysis branch of algebra and is defined as:

$$K^{\#} = (\lambda_1/\lambda_p)^{1/2} = \max(\text{singular value})/\min(\text{singular value}). \quad (12.8)$$

Note that $K^{\#}$ of $X'X$ is infinitely large when $\lambda_p = 0$, or the matrix is singular (non-invertible). The reader may wonder why the definition Eq. (12.8) has λ_1 in the numerator, since for singularity only λ_p matters. The answer is that the presence of λ_1 makes the ratio not sensitive to units of measurement of various columns of X. If columns of X are perpendicular (orthogonal) to each other, $\lambda_1 = \lambda_p = 1$ and $K^{\#}$ is unity, this is the opposite of collinearity. Thus, $K^{\#} \in [1, \infty)$.

It is proved in numerical analysis literature that when $K^{\#}$ is large, the effect of small perturbation in X can be large for OLS coefficients, Vinod and Ullah (1981, p. 128). It is generally agreed that one should avoid ill-conditioned matrices, but the magnitude of $K^{\#}$ to determine when a particular matrix is seriously ill conditioned depends on the subject matter under study. We have computed the condition number (= 5.515602) for the numerical example of the previous section using R.

12.3.1   *Rule of Thumb for a Large Condition Number*

In my experience with economic data, if $K^{\#} > 10p$, (p = number of regressors) it is safe to conclude that ill conditioning is serious enough to require remedial action.

The $K^{\#} = 5.515602$ for the numerical example is not too large according to this rule of thumb and we can conclude that the artificial data is not seriously ill-conditioned.

For another example, if one column of X had the GDP dollars (say

$956,324,336,000)$ and another column of X has unemployment rate (say 2, 7, 9 as percentages), this can create ill-conditioning. Sometimes simple change in units of measurement so that variable like the GDP are reported in hundreds of billions of dollars making GDP comparable to unemployment and interest rates is helpful in avoiding ill-conditioning. If this does not help, one might need a more drastic action such as 'ridge regression' discussed in Vinod and Ullah (1981) and Vinod (2008a).

12.3.2　*Pascal Matrix is Ill-conditioned*

A Pascal triangle is well known for finding the Binomial coefficients. The Pascal matrix is created by having numbers of a Pascal Triangle in the top left corner of its matrix. The R package 'matrixcalc' has a function for computing the Pascal matrix of any dimension.

```
# R program snippet 12.3.2.1 is next.
library(matrixcalc)
cond.no=rep(NA,10)
for (n in 5:14){
A=pascal.matrix(n);
if (n==5) print(A)
ei=eigen(A)
cond.no[n-4]=sqrt(ei$val[1]/ei$val[n])
if (n==5) print(ei$val/eigen(solve(A))$values)}
cond.no
```

The output prints only the 5×5 Pascal matrix for brevity. We do however print their $K^{\#}$ condition numbers (defined in Eq. (12.8)) as n increases from 5 to 14. They show that Pascal matrix A is ill-conditioned but not singular. An interesting property of Pascal matrices is that the eigenvalues of A and A^{-1} are identical. This is verified in the snippet 12.3.2.1 by the command 'if (n==5) print(ei*val/eigen(solve(A))*values)' which computes the ratios of all eigenvalues and their inverses computed in R by 'solve(A)'. The output shows that the ratio is 1 for all 5 eigenvalues.

```
> for (n in 5:14){
+ A=pascal.matrix(n);
+ if (n==5) print(A)
+ ei=eigen(A)
+ cond.no[n-4]=sqrt(ei$val[1]/ei$val[n]) }
```

```
      [,1] [,2] [,3] [,4] [,5]
[1,]    1    1    1    1    1
[2,]    1    2    3    4    5
[3,]    1    3    6   10   15
[4,]    1    4   10   20   35
[5,]    1    5   15   35   70     .
> cond.no
 [1] 9.229043e+01 3.328463e+02 1.222065e+03 4.543696e+03
     1.705235e+04
 [6] 6.446089e+04 2.450786e+05 9.361588e+05 3.589890e+06
     1.381161e+07
```

The above output lists condition numbers for various Pascal matrices as $n = 5, 6, \ldots, 14$. Note that when $n = 9, K^{\#} = 17052.35$, already a huge number almost 1895 times n, much worse than the 10 times suggested by the rule of thumb in Sec. 12.3.1.

12.4 Hilbert Matrix is Ill-conditioned

In testing numerical accuracy and software code researchers often use the Hilbert matrix. If B is a Hilbert matrix, its entries are fractions depending on the row and column numbers:

$$B_{ij} = \frac{1}{i+j-1}. \tag{12.9}$$

For example, a 3×3 Hilbert matrix is given by

$$B_n = \begin{bmatrix} 1 & \frac{1}{2} & \frac{1}{3} \\ \frac{1}{2} & \frac{1}{3} & \frac{1}{4} \\ \frac{1}{3} & \frac{1}{4} & \frac{1}{5} \end{bmatrix}. \tag{12.10}$$

Note that $n \times n$ Hilbert matrices are square, symmetric and positive definite, similar to those discussed in Sec. 11.3. They arise from the integral $\int_0^1 x^{i+j-2} dx$, or as a Gramian matrix for (polynomial) powers of x. Starting with a set of vectors in an inner product space, Hermitian matrix of inner products is known as the Gramian matrix.

More important, Hilbert matrices are ill-conditioned. The R package 'Matrix' creates Hilbert matrices by the function 'Hilbert' as in the following snippet.

R program snippet **12.4.1** is next.

```
ei=eigen(Hilbert(3)); sqrt(ei$val[1]/ei$val[3])
ei=eigen(Hilbert(4)); sqrt(ei$val[1]/ei$val[4])
ei=eigen(Hilbert(5)); sqrt(ei$val[1]/ei$val[5])
ei=eigen(Hilbert(6)); sqrt(ei$val[1]/ei$val[6])
ei=eigen(Hilbert(13)); sqrt(ei$val[1]/ei$val[13])
ei=eigen(Hilbert(14)); sqrt(ei$val[1]/ei$val[14])
ei$val[14]
```

The output (suppressed for brevity) of the snippet 12.4.1 shows that the condition number of Hilbert matrices of dimensions $n = 3, \ldots, 6$ are respectively: $K^{\#}(B_n) = (22.89229, 124.5542, 690.3675, 3866.66)$. The output from the last three lines of the snippet is:

```
> ei=eigen(Hilbert(13)); sqrt(ei$val[1]/ei$val[13])
[1] 767122660
> ei=eigen(Hilbert(14)); sqrt(ei$val[1]/ei$val[14])
[1] NaN
Warning message:
In sqrt(ei$val[1]/ei$val[14]) : NaNs produced
> ei$val[14]
[1] -5.408053e-18
```

Hilbert matrices are ill-conditioned in the sense that their condition numbers are rising very fast as n increases. In fact, R software for eigenvalue computation beaks down when $n = 14$, since it cannot compute all positive eigenvalues, even though we know from theory that eigenvalues are real and positive. R incorrectly computes the $\lambda_{14}(B_{14}) = -5.408053e - 18$, a negative number near zero, but correctly warns that dividing by such a small near zero number is 'NaN' or infinity or not a number. This shows that even good software does not have infinite numerical accuracy. We discuss numerical accuracy in Chapter 17.

## Chapter 13

Simultaneous Reduction and Vec Stacking

Section 9.5 discusses the eigenvalue-eigenvector decomposition of A as $G\Lambda G'$, which has reduced one square matrix A into a diagonal form Λ. In Statistical hypothesis testing of one-way classification model we have ratios of two quadratic forms where simultaneous reduction of the numerator and denominator matrix is needed. This chapter considers such situations. Also, we may want to know when the effect of G which diagonalizes A on another matrix B is to diagonalize B also. This leads us to consider commuting matrices.

Later in the chapter we consider the 'vec' (vectorization) operator in conjunction with the Kronecker product. We had encountered the 'vec' operator in Sec. 11.6.1. I find that if a matrix theory claim fails to be supported by numerical evaluations using R, it is usually due to typing errors or misprints. Merely staring at formulas or viewing the proofs is far more time consuming (less fun) than checking the results in R after using randomly chosen matrices.

13.1 Simultaneous Reduction of Two Matrices to a Diagonal Form

Consider a pair of real $n \times n$ symmetric matrices A, B, of which B is positive definite. Denote by Λ a diagonal matrix with eigenvalues λ_i of $B^{-1}A$ along its diagonal. Then there exists a nonsingular matrix C such that

$$A = (C^{-1})'\Lambda C^{-1}, \quad B = (C^{-1})'C^{-1}, \qquad (13.1)$$

where Λ is a diagonal matrix. If we write the i-th column of C^{-1} as h_i we also have:

$$A = \lambda_1 h_1 h_1' + \lambda_2 h_2 h_2' + \ldots + \lambda_n h_n h_n',$$
$$B \quad = h_1 h_1' + h_2 h_2' + \ldots + h_n h_n'. \tag{13.2}$$

An implication of Eq. (13.2) is that there exists a transformation from x to y defined by $y = C^{-1}x$ such that two quadratic forms $x'Ax$ and $x'Bx$ are simultaneously reduced to simpler forms containing square terms y_i^2 only. In other words,

$$x'Ax = \lambda_1 y_1^2 + \lambda_2 y_2^2 + \ldots + \lambda_n y_n^2,$$
$$x'Bx \quad = y_1^2 + y_2^2 + \ldots + y_n^2. \tag{13.3}$$

Let G be an $n \times n$ orthogonal matrix of eigenvectors in Sec. 9.5, which reduces the $n \times n$ symmetric matrix A to the diagonal form, in the sense that $A = G'\Lambda G$ can be re-written upon pre-multiplication by G and post-multiplication by G' as:

$$GAG' = \Lambda, \tag{13.4}$$

where the diagonalization is visible on the right hand side.

Now, $AB = BA$ is a necessary and sufficient condition that G also reduces another $n \times n$ symmetric matrix B to a diagonal form with the eigenvalues of B denoted by ϕ_i along the diagonal of Φ, in the sense that

$$GBG' = \Phi, \tag{13.5}$$

where $\Phi = diag(\phi_i,$ and $i = 1, 2, \ldots, n)$.

A practical question regarding two quadratic forms involving A, B matrices in this context is whether or not both can be reduced to simpler forms as in Eq. (13.3) by a single orthogonal transformation. The answer is true if $(AB = BA)$. Regarding Eq. (13.4) and Eq. (13.5) matrix theory suggests that if and only if A and B commute, there is an orthogonal matrix G with common eigenvectors which simultaneously diagonalizes both matrices.

13.2 Commuting Matrices

Commonly encountered commuting matrices are two diagonal matrices. If A and B are diagonal matrices, row-column multiplication $AB = BA$ gives product of the two diagonals along the diagonal. Are there any other commuting matrices? Since commuting $(AB = BA)$ is both necessary and sufficient condition for simultaneous reduction of Sec. 13.1, we propose to use simultaneous reduction to visualize commuting matrices in R. Since matrix theory books do not seem to give non-singular examples of commuting matrices, our hands-on approach demands that we construct an example using R.

The idea for the following snippet is to start with a simple 3×3 matrix, find its eigenvalues and eigenvector decomposition $A = G\Lambda G'$. Now change the eigenvalues (elements of Λ matrix) a bit as elements of $\tilde{\Lambda}$ but keep the same eigenvectors and reconstruct $B = G\tilde{\Lambda}G'$ matrix from the known decomposition and check if $AB = BA$ holds for this artificial B.

```
# R program snippet 13.2.1 is next.
#R.snippet construct commuting B from A
set.seed(641);a=sample(11:99)
A=matrix(a[1:9],3);A
solve(A)# is it nonsingular
ea=eigen(A)
G=ea$vec
D=ea$val;D #original eigenvalues
D2=c(20, 5,9) #arbitrary new eigenvalues
Lam=diag(D)
G %*% Lam %*% t(G) #should equal original A
Lam2=diag(D2)
B=G %*% Lam2 %*% t(G); B #define B
round(A%*%B-B%*%A,7)#should have all zeros
```

The following output shows that matrix theory claim does not hold for this simple 3×3 matrix.

```
> A=matrix(a[1:9],3);A
     [,1] [,2] [,3]
[1,]   89   68   48
[2,]   19   50   21
[3,]   94   59   17
> solve(A)# is it nonsingular
              [,1]         [,2]         [,3]
[1,]   0.004131924  -0.01780233   0.01032450
[2,]  -0.017536778   0.03185512   0.01016517
[3,]   0.038015827  -0.01211960  -0.03354400
> ea=eigen(A)
```

```
> G=ea$vec
> D=ea$val;D #original eigenvalues
[1] 153.50291  26.04511 -23.54802
> D2=c(20, 5,9) #arbitrary new eigenvalues
> Lam=diag(D)
> G %*% Lam %*% t(G) #should equal original A
            [,1]      [,2]      [,3]
[1,] 88.70519 18.19119 82.05749
[2,] 18.19119 24.59942 21.38310
[3,] 82.05749 21.38310 42.69538
> Lam2=diag(D2)
> B=G %*% Lam2 %*% t(G); B #define B
            [,1]         [,2]         [,3]
[1,] 12.946921 2.40863970  7.81838583
[2,]  2.408640 4.58909117  0.08713453
[3,]  7.818386 0.08713453 16.46398764
> round(A%*%B-B%*%A,7)#should have all zeros
            [,1]          [,2]       [,3]
[1,] -241.6224   -931.4978  687.0867
[2,]  220.8573   -121.3345  285.1829
[3,] -753.0738  -1008.7324  362.9569
```

The snippet shows that the eigenvalue-eigenvector decomposition $A = G\Lambda G'$ does not hold, since we cannot get back the original A. It is therefore not surprising that $AB - BA$ is not at all a matrix of all zeros.

Let us change notation and denote the above A as A^0, construct a new and modified $A = (1/2)(A^0 + t(A^0))$ forcing it to be symmetric. We note that the matrix theory assertion does hold for the modified case in the following snippet.

\# R program snippet **13.2.2** Creating Symmetric matrix.

```
set.seed(641);a=sample(11:99)
A0=matrix(a[1:9],3) #old A is now A0
A=(A0+t(A0))/2; A #view new A
ea=eigen(A)
G=ea$vec
D=ea$val;D #original eigenvalues
Lam=diag(D)
G %*% Lam %*% t(G) #should equal original A2
```

```
D2=c(20, 5,9) #arbitrary new eigenvalues
Lam2=diag(D2); Lam2#view new diagonal
B=G %*% Lam2 %*% t(G); B #define B
round(A%*%B-B%*%A,7)#should have all zeros
```

Note that the trick of adding a transpose and dividing by 2 to construct a symmetric matrix does work. The new A is indeed visually symmetric (entries above and below diagonal match). The decomposition does indeed get back the original A. Arbitrary choice of new eigenvalues do yield a new B (constructed by changing the eigenvalues while keeping the eigenvectors unchanged) which is symmetric and does commute with A.

```
> A=(A0+t(A0))/2; A #view new A
      [,1] [,2] [,3]
[1,] 89.0 43.5   71
[2,] 43.5 50.0   40
[3,] 71.0 40.0   17
> ea=eigen(A)
> G=ea$vec
> D=ea$val;D #original eigenvalues
[1] 162.64942  22.16604 -28.81547
> Lam=diag(D)
> G %*% Lam %*% t(G) #should equal original A2
      [,1] [,2] [,3]
[1,] 89.0 43.5   71
[2,] 43.5 50.0   40
[3,] 71.0 40.0   17
> D2=c(20, 5,9) #arbitrary new eigenvalues
> Lam2=diag(D2); Lam2#view new diagonal
     [,1] [,2] [,3]
[1,]   20    0    0
[2,]    0    5    0
[3,]    0    0    9
> B=G %*% Lam2 %*% t(G); B #define B
           [,1]      [,2]       [,3]
[1,] 14.081666 5.464140   3.850553
[2,]  5.464140 8.317109   2.699703
[3,]  3.850553 2.699703 11.601225
> round(A%*%B-B%*%A,7)#should have all zeros
      [,1] [,2] [,3]
```

```
[1,]    0    0    0
[2,]    0    0    0
[3,]    0    0    0
```

In the above reduction to a diagonal form we insisted on G being an orthogonal matrix, with common eigenvectors. For some purposes it is not necessary to have an orthogonal matrix, but one would like to relax the commutativity requirement. Then matrix theory has the following result:

Given that A is a real symmetric, nonsingular and positive definite matrix, and that B is a real symmetric matrix, there exists a non-singular matrix S which diagonalizes A to the identity matrix and simultaneously diagonalizes B, such that

$$S'AS = I \quad \text{and} \quad S'BS = \Phi, \tag{13.6}$$

where $\Phi = diag(\phi_i, \text{ and } i = 1, 2, \cdots, n)$. How to find such S matrix? An answer is proposed in the following snippet.

```
# R program snippet 13.2.3 Simultaneous Diagonalization.
A=matrix(c(25,4,4,16),2);A
B=matrix(c(9,2,2,4),2);B
ea=eigen(A)
Lam=diag(ea$value)
G=ea$vector
Laminv=sqrt(solve(Lam))
S=G%*% Laminv
#S=Laminv%*%G  fails
t(S)%*%A%*% S #should be identity
t(S)%*% B %*% S#should be diagonal
```

We use square roots of eigenvalues to construct a choice of $S = G\Lambda^{-1/2}$ matrix appearing in Eq. (13.6) in the following snippet.

```
> A=matrix(c(25,4,4,16),2);A
     [,1] [,2]
[1,]   25    4
[2,]    4   16
> B=matrix(c(9,2,2,4),2);B
     [,1] [,2]
[1,]    9    2
[2,]    2    4
```

```
> ea=eigen(A)
> Lam=diag(ea$value)
> G=ea$vector
> Laminv=sqrt(solve(Lam))
> S=G%*% Laminv
> #S=Laminv%*%G  fails
> t(S)%*%A%*% S #should be identity
              [,1]            [,2]
[1,] 1.000000e+00 -9.722583e-17
[2,] 2.329679e-17  1.000000e+00
> t(S)%*% B %*% S#should be diagonal
              [,1]            [,2]
[1,]  0.365647038 -0.008475794
[2,] -0.008475794  0.228102962
```

Considering the last two matrices in the output of the snippet 13.2.3 Equation (13.6) seems to work, except for numerical rounding and truncation errors. Note that $S'AS$ is close to the identity matrix and $S'BS$ is close to a diagonal matrix. If instead we post-multiply by G and choose $S = \Lambda^{-1/2}G$, we do not get close to identity or diagonal matrices.

13.3 Converting Matrices Into (Long) Vectors

Convention: Let A be a $T \times m$ matrix; the notation $vec(A)$ will mean the Tm-element column vector whose first set of T elements are the first column of A, that is $a_{.1}$ using the dot notation for columns; the second set of T elements are those of the second column of A, $a_{.2}$, and so on. Thus $A = [a_{.1}, a_{.2}, \cdots, a_{.m}]$ in the dot notation.

For example, if

$$A = \begin{bmatrix} a_{11} & a_{12} \\ a_{21} & a_{22} \end{bmatrix}, \text{ then } \quad vec(A) = \begin{bmatrix} a_{11} \\ a_{21} \\ a_{12} \\ a_{22} \end{bmatrix}, \tag{13.7}$$

where one simply stacks the columns one below the other in a long column vector.

In R package called 'fUtilities,' Wuertz and many others (2010), there is a function called 'vec' which does the vec conversion of a matrix into a long vector after stacking it column-wise. Actually, R stores its $n \times m$ matrices

A in the 'vec' notation as it were. Hence one can get the long column vector vec(A) by simply using the R command 'matrix(A,n*m,1)'.

```
# R program snippet 13.2.4 is next.
library(fUtilities)
A=matrix(1:4,2); A #A as 2 by 2 matrix
vec(A)# vec operation on A
```

The stacking is seen next

```
> A=matrix(1:4,2); A #A as 2 by 2 matrix
     [,1] [,2]
[1,]   1    3
[2,]   2    4
> vec(A)# vec operation on A
     [,1]
[1,]   1
[2,]   2
[3,]   3
[4,]   4
```

A consequence of the above convention is that $Vec(AB)$ (Vec of a product of two matrices) can be written as an expression involving a Kronecker product, the identity matrix, $Vec(B)$ and $Vec(A)$ as shown in Eq. (13.8) below.

Let A, B be $T \times m$, $m \times q$ respectively. Then using the Kronecker product notation, we have

$$vec(AB) = (B' \otimes I)vec(A) = (I \otimes A)vec(B) \qquad (13.8)$$

For example, if $A = a_{ij}$ and $B = b_{ij}$ are two 2×2 matrices, then row-column multiplication yields

$$AB = \begin{bmatrix} a_1b_1 + a_2b_3 & a_1b_2 + a_2b_4 \\ a_3b_1 + a_4b_3 & a_3b_2 + a_4b_4 \end{bmatrix} \qquad (13.9)$$

It is useful to check this in R.

```
# R program snippet 13.2.5 is next.
A=matrix(1:4,2);A
B=matrix(11:14,2);B
LHS=vec(A%*%B) #stacked AB
```

```
myi=diag(2)#identity matrix
RHS1=(myi %x% A) %*% vec(B)
RHS2=(t(B) %x% myi) %*% vec(A)
cbind(LHS,RHS1,RHS2) #should be equal
```

The left side of Eq. (13.8) is estimated in R by the command 'vec(A%*%B)' and the two right sides by the next two lines of the snippet 13.2.5 marked RHS1 and RHS2.

```
> A=matrix(1:4,2);A
     [,1] [,2]
[1,]   1    3
[2,]   2    4
> B=matrix(11:14,2);B
     [,1] [,2]
[1,]  11   13
[2,]  12   14
> LHS=vec(A%*%B) #stacked AB
> myi=diag(2)#identity matrix
> RHS1=(myi %x% A) %*% vec(B)
> RHS2=(t(B) %x% myi) %*% vec(A)
> cbind(LHS,RHS1,RHS2) #should be equal
     [,1] [,2] [,3]
[1,]  47   47   47
[2,]  70   70   70
[3,]  55   55   55
[4,]  82   82   82
```

We use the 'cbind' function of R to report the left hand side and two right hand sides side-by-side to quickly note that they are identical. This makes Eq. (13.8) plausible with the help of a simple example.

13.3.1 *Vec of ABC*

Vec of a product of three matrices satisfies the following rule involving vec of the middle matrix. Let A, B, C be conformably dimensioned matrices. Then the vec operator on their product satisfies:

$$vec(ABC) = (C' \otimes A)\, vec(B) \tag{13.10}$$

Now we use R to check the proposition at the start of this subsection. We use the 'cbind' function of R to report the left and right hand sides in the following snippet.

```
# R program snippet 13.2.6 is next.
A=matrix(1:4,2);A
B=matrix(11:14,2);B
C=-matrix(21:24,2); C #view C
LHS=vec(A %*% B %*% C)
RHS=(t(C) %x% A)%*% vec(B)
cbind(LHS,RHS)
```

We do not reproduce the part of the output printing A and B matrices.

```
> C=-matrix(21:24,2); C #view C
      [,1] [,2]
[1,]   -21   -23
[2,]   -22   -24
> LHS=vec(A %*% B %*% C)
> RHS=(t(C) %x% A)%*% vec(B)
> cbind(LHS,RHS)
        [,1]   [,2]
[1,] -2197 -2197
[2,] -3274 -3274
[3,] -2401 -2401
[4,] -3578 -3578
```

Clearly the left and right sides match and Eq. (13.10) is supported.

The $vec(ABC)$ needs the Kronecker product with Identity, if we want to avoid $vec(B)$ appearing in Eq. (13.10).

Let A_1, A_2, A_3 be conformably dimensioned matrices. Then

$$vec(A_1 A_2 A_3) = (I \otimes A_1 A_2)\, vec(A_3) \qquad (13.11)$$
$$= (A_3' \otimes A_1)\, vec(A_2)$$
$$= (A_3' A_2' \otimes I)\, vec(A_1).$$

For a hands-on check, let us keep the two matrices from the previous snippet in the memory of R and only include C matrix with all negative values (for fun).

```
# R program snippet 13.2.7 is next.
```

```
#R.snippet recall A and B before in memory
A=matrix(1:4,2);myi=diag(2)
B=matrix(11:14,2)
C=-matrix(21:24,2)
LHS=vec(A %*% B %*% C)
RHS1=(myi %x% (A%*%B)) %*% vec(C)
RHS2=(t(C) %x% A) %*% vec(B)
RHS3=((t(C)%*%t(B)) %x% myi) %*% vec(A)
cbind(LHS,RHS1,RHS2,RHS3)
```

The left and right sides are denoted in standard notation and we check that they are all equal.

```
> cbind(LHS,RHS1,RHS2,RHS3)
      [,1]  [,2]  [,3]  [,4]
[1,] -2197 -2197 -2197 -2197
[2,] -3274 -3274 -3274 -3274
[3,] -2401 -2401 -2401 -2401
[4,] -3578 -3578 -3578 -3578
```

Thus, despite its rather strange appearance, the matrix theory result Eq. (13.11) does hold true.

13.3.2 Vec of $(A + B)$

Vec of a Sum of Two Matrices is the sum of vecs. Let A, B be $T \times n$. Then

$$vec(A + B) = vec(A) + vec(B). \tag{13.12}$$

Corollary: Let A, B, C, D be conformably dimensioned matrices. Then the two propositions together imply the following result.

$$
\begin{aligned}
vec[(A + B)(C + D)] &= [(I \otimes A) + (I \otimes B)][vec(C) + vec(D)] \\
&= [(C' \otimes I) + (D' \otimes I)][vec(A) + vec(B)]
\end{aligned} \tag{13.13}
$$

For a hands-on check, let us keep the two matrices from the previous snippet in the memory of R and only include C matrix with all negative values (for fun).

```
# R program snippet 13.2.7 is next.
#R.snippet recall I,A,B and C before in memory
library(fUtilities)
```

```
D=-matrix(31:34,2); D #view D
LHS=vec((A+B) %*% (C+D))
RHS1=((myi %x% A)+(myi%x%B)) %*% (vec(C)+vec(D))
RHS2=((t(C) %x% myi)+(t(D)%x%myi)) %*% (vec(A)+vec(B))
cbind(LHS,RHS1,RHS2)
```

```
> D=-matrix(31:34,2); D #view D
     [,1] [,2]
[1,]  -31  -33
[2,]  -32  -34
> LHS=vec((A+B) %*% (C+D))
> RHS1=((myi %x% A)+(myi%x%B)) %*% (vec(C)+vec(D))
> RHS2=((t(C) %x% myi)+(t(D)%x%myi)) %*% (vec(A)+vec(B))
> cbind(LHS,RHS1,RHS2)
        [,1]   [,2]   [,3]
[1,] -1488 -1488 -1488
[2,] -1700 -1700 -1700
[3,] -1600 -1600 -1600
[4,] -1828 -1828 -1828
```

Equation (13.13) clearly does check out.

13.3.3 Trace of AB In Terms of Vec

Trace of a Product of Two Matrices in the Vec notation (prime of the vec of a prime). Let A, B be conformably dimensioned matrices. Then

$$tr(AB) = vec(A')'vec(B) = vec(B')'vec(A). \qquad (13.14)$$

We can check Eq. (13.14) by using R in the following snippet.

```
# R program snippet 13.2.8 is next.
LHS=sum(diag(A %*%B))
RHS1=t(vec(t(A))) %*% vec(B)
RHS2=t(vec(t(B))) %*% vec(A)
cbind(LHS,RHS1,RHS2)
```

The trace is defined from the sum of diagonals and two right hand sides are computed.

```
> LHS=sum(diag(A %*%B))
> RHS1=t(vec(t(A))) %*% vec(B)
> RHS2=t(vec(t(B))) %*% vec(A)
> cbind(LHS,RHS1,RHS2)
     LHS
[1,] 129 129 129
```

Equation (13.14) is clearly supported by the example.

13.3.4 *Trace of ABC In Terms of Vec*

Trace of a product of three matrices involves the prime of a vec and a Kronecker product with the identity. Let A_1, A_2, A_3 be conformably dimensioned matrices. Then matrix theory claims the following:

$$tr(A_1 A_2 A_3) = vec(A_1')'(A_3' \otimes I)vec(A_2) \qquad (13.15)$$
$$= vec(A_1')'(I \otimes A_2)vec(A_3)$$
$$= vec(A_2')'(I \otimes A_3)vec(A_1)$$
$$= vec(A_2')'(A_1' \otimes I)vec(A_3)$$
$$= vec(A_3)'(A_2' \otimes I)vec(A_1)$$

Now we check Eq. (13.15) in R.

```
# R program snippet 13.2.9 using vec operation.
library(fUtilities) #to get vec to work
set.seed(9754);a=sample(1:100)
A1=matrix(a[1:9],3);A1# view A1
A2=matrix(a[11:19],3);A2# view A2
A3=matrix(a[21:29],3);A3# view A3
LHS=sum(diag(A1%*%A2%*%A3));LHS
myi=diag(3)#3 by 3 identity matrix
RHS1=t(vec(t(A1)))%*%(t(A3) %x% myi) %*% vec(A2)
RHS2=t(vec(t(A1)))%*%(myi %x% A2) %*% vec(A3)
RHS3=t(vec(t(A2)))%*%(myi %x% A3) %*% vec(A1)
RHS4=t(vec(t(A2)))%*%(t(A1) %x% myi) %*% vec(A3)
RHS5=t(vec(t(A3)))%*%(t(A2) %x% myi) %*% vec(A1)
cbind(LHS,RHS1,RHS2,RHS3,RHS4,RHS5)
```

The last line prints all left and right sides together. We use a random sampling out of the first 100 numbers to construct our matrices A1 to A3 in R.

```
> set.seed(9754);a=sample(1:100)
> A1=matrix(a[1:9],3);A1# view A1
     [,1] [,2] [,3]
[1,]   80   18   66
[2,]   96   61   13
[3,]   79    5    8
> A2=matrix(a[11:19],3);A2# view A2
     [,1] [,2] [,3]
[1,]   52   39   98
[2,]   46   23   90
[3,]   87   21   27
> A3=matrix(a[21:29],3);A3# view A3
     [,1] [,2] [,3]
[1,]   36   31   54
[2,]   81   91   49
[3,]   45    9   65
> LHS=sum(diag(A1%*%A2%*%A3));LHS
[1] 3181142
> myi=diag(3)#3 by 3 identity matrix
> RHS1=t(vec(t(A1)))%*%(t(A3) %x% myi) %*% vec(A2)
> RHS2=t(vec(t(A1)))%*%(myi %x% A2) %*% vec(A3)
> RHS3=t(vec(t(A2)))%*%(myi %x% A3) %*% vec(A1)
> RHS4=t(vec(t(A2)))%*%(t(A1) %x% myi) %*% vec(A3)
> RHS5=t(vec(t(A3)))%*%(t(A2) %x% myi) %*% vec(A1)
> cbind(LHS,RHS1,RHS2,RHS3,RHS4,RHS5)
          LHS
[1,] 3181142 3181142 3181142 3181142 3181142 3181142
```

This shows almost magically that all five right hand sides match precisely and the claim in Eq. (13.15) is supported for the example at hand.

Now we consider the following result regarding trace of a product of four matrices.

$$tr(ABCD) = (vec(D'))'(C' \otimes A)(vec(B)) \qquad (13.16)$$
$$= (vec(D))'(A \otimes C')(vec(B'))$$

```
# R program snippet 13.2.10 trace properties.
#R.snippet  bring A1 to A3 in memory
A4=matrix(a[31:39],3);A4# view A4
```

```
LHS=sum(diag(A1%*%A2%*%A3%*%A4))
RHS1=t(vec(t(A4)))%*%(t(A3) %x% A1) %*% vec(A2)
RHS2=t(vec(A4))%*%(A1 %x% t(A3)) %*% vec(t(A2))
cbind(LHS,RHS1,RHS2)
```

The claim Eq. (13.16) is supported by the example. We skipped a few lines in reproducing the output for brevity.

```
> A4=matrix(a[31:39],3);A4# view A4
     [,1] [,2] [,3]
[1,]   15   48   73
[2,]   82   89   34
[3,]   38   63   86
> cbind(LHS,RHS1,RHS2)
           LHS
[1,] 573913667 573913667 573913667
```

If one misses a transpose in any of these expressions, the results are found to be very different. Our hands-on approach shows that possible misprints in matrix theory are revealed by using R to validate the claims.

13.4   Vech for Symmetric Matrices

The 'vec' operator stacks all elements column-wise. What if one wants to stack only the lower triangular part of the matrix?. The 'vech' operator is available for this purpose and offers some economy of storage. R package 'matrixcalc' has the function 'vech' for this purpose. We illustrate its use in the following snippet. select columns of lower triangular part of a matrix.

```
# R program snippet 13.4.1 is next.
library(matrixcalc)
A=matrix(1:9,3);A #print A
t(vec(A))# print transpose of vec of A
t(vech(A))#print transpose of vech of A
```

The output of snippet 13.4.1 shows that it is economical to use 'vech' instead of 'vec' if the matrix is symmetric, but not otherwise.

```
> A=matrix(1:9,3);A #print A
     [,1] [,2] [,3]
[1,]    1    4    7
```

```
[2,]    2    5    8
[3,]    3    6    9
> t(vec(A))# print transpose of vec of A
     [,1] [,2] [,3] [,4] [,5] [,6] [,7] [,8] [,9]
[1,]    1    2    3    4    5    6    7    8    9
> t(vech(A))#print transpose of vech of A
     [,1] [,2] [,3] [,4] [,5] [,6]
[1,]    1    2    3    5    6    9
```

In the example above 'vech' loses information about the elements in the upper triangle of the matrix. This should serve as a warning against misuse of 'vech' if one wants 'vec'.

Chapter 14

Vector and Matrix Differentiation

Calculus provides important tools for a study of functions including transformations, approximations, minimization, maximization, relative changes, etc. When functions involve several variables, we need tools combining calculus and matrix algebra. This chapter contains a series of formulas for matrix differentiation with respect to (wrt) a scalar, vector or a matrix. Unfortunately, hands-on experience for these formulas is limited using R, even though R does have symbolic manipulation to compute derivatives. There is a need for an R package which will help check these formulas in a convenient fashion. The package 'matrixcalc' by Novomestky (2008) based on Magnus and Neudecker (1999) is a good start.

14.1 Basics of Vector and Matrix Differentiation

Let us begin with a familiar minimization problem from the derivation of the least squares regression considered in Sec. 8.5 to motivate this chapter. We minimize the error sum of squares by using the first order (necessary) conditions for a minimum. In Sec. 8.9 we differentiate the error sum of squares ($\epsilon'\epsilon$, defined from a vector ϵ of errors) with respect to (wrt) the vector of regression parameters β. The 'normal equations' containing matrices and vectors are obtained by setting this derivative equal to zero in Eq. (8.32). Many practical problem of this type can be conveniently solved if one learns to compute derivatives of vectors and matrices wrt vectors and matrices. No fundamentally new calculus concepts are involved in carrying out such matrix differentiations.

Matrix differentiation is a powerful tool for research by allowing us to quickly write valid answers to various calculus problems involving several variables at the same time. Apart from derivatives of sums of squares, this

chapter will consider derivatives of quantities like $tr(AX)$ wrt the elements
of X, and derivatives of quantities like $Ax, z'Ax$ with respect to the elements
of (the vectors) x and/or z.

Symbolic differentiation or derivatives in R are cumbersome, but possible. Derivative of a vector containing $(\cos(x), \sin(x), \cos^2(x), x^3)$ wrt x
is computed in the following snippet. If 'x' is a numerical R object the
function 'cos(x)' will compute the cosines of the numbers. R needs to distinguish between such computation and symbolic manipulation. This is
accomplished by defining an object as an 'expression'. R function 'D' for
derivatives computes the derivative of the expression object wrt 'x', say as
follows. The R command 'D(expression(x²), "x")' computes the derivative
of x^2 wrt x as 2x, where R expects double quotation marks around 'x' to
distinguish it from the usual R object, making it a character object wrt
which to differentiate. Unfortunately, if we use 'D(.,.)' command on a vector of expression values, R computes the symbolic derivative of only the
first expression. Hence one must use the R function called 'lapply' which
applies the function D to the 'list' in the 'expression' vector.

```
#R.snippet derivative of a vector wrt scalar
D(expression(x^2),"x")
y=expression(sin(x), cos(x), cos(x)^2, x^3)
dydx=lapply(y,D,"x");dydx
```

R correctly computes a vector of derivatives.

```
> D(expression(x^2),"x")
2 * x
> y=expression(sin(x), cos(x), cos(x)^2, x^3)
> dydx=lapply(y,D,"x");dydx
[[1]]
cos(x)
[[2]]
-sin(x)
[[3]]
-(2 * (sin(x) * cos(x)))
[[4]]
3 * x^2
```

**Convention of using the same numerator subscript along each
row.** The reader will find that a good grasp of the following (arbitrary)

convention is essential for understanding matrix derivatives.

Let $y = \psi(x)$, where y is a $T \times 1$ vector and x is an $n \times 1$ vector. The symbol

$$\frac{\partial y}{\partial x} = \{\frac{\partial y_i}{\partial x_j}\}, \quad i = 1, 2, \cdots, T, \quad \text{and} \quad j = 1, 2 \cdots, n, \quad (14.1)$$

denotes the $T \times n$ matrix of first-order partial derivatives (gradients). We provide an examples of the convention in Eq. (14.3) and discuss it following that equation.

In Statistics, $(\partial y / \partial x)$ is called the Jacobian matrix of the transformation from x to y. Consider the density $f(x)$ of a random variable x. Now assume that there is another random variable with density $g(y)$, where the transformation from x to y is '1 to 1' such that $x = u^{-1}(y)$. If $|J|$ is the absolute value of the determinant of the matrix $(\partial u^{-1}(y)/\partial y)$, similar to the left hand side of Eq. (14.1), we have: $g(y) = f(x)|J|$, as the relation linking the two densities.

It is useful to consider a univariate example where the Jacobian $|J|$ is simply a scalar to fix the ideas. Consider $f(x) = [1/\sqrt{(2\pi)}]exp(-x^2/2)$ as the standard Normal density $N(0,1)$, spelled out for the general case: $N(\mu, \sigma^2)$, in Eq. (15.1). Now, $g(y)$ is the Lognormal density linked by $y = exp(x)$. We shall now derive the Lognormal density by using a Jacobian of transformation formula $g(y) = f(x)|J|$. Note that $x = log(y), dx/dy = (1/y)$ implies that $|1/y|$ is the (univariate) scalar Jacobian of transformation, defined only when $y > 0$. Hence we have:

$$g(y) = [1/y\sqrt{(2\pi)}]exp[-(log(x)^2/2)], \quad \text{if} \quad y > 0,$$
$$= 0, \quad \text{otherwise.} \quad (14.2)$$

Note that Eq. (14.2) is in the form of the standard Lognormal density. Other interesting examples using the Jacobian matrix are: (i) construction of truncated random variable, and (ii) multivariate distributions in the context of 'seemingly unrelated regressions.'

The R package 'rootSolve,' Soetaert and Herman (2009), has functions called 'jacobian.full' and 'jacobian.band' which returns a forward difference approximation of the jacobian for full and banded 'ordinary differential equation' (ODE) problems. The R package 'numDeriv,' Gilbert (2009) has a function 'jacobian' to compute the m by n numerical approximation of the gradient of a real m-vector valued function with n-vector argument. For example, if we use a vector of two items defined in R by 'c(six(x), cos(x))' its Jacobian will be computed in the following snippet.

R program snippet **14.3.1** is next.

```
library(numDeriv)
func2 <- function(x) c(sin(x), cos(x))
x <- (0:1)*2*pi; x
jacobian(func2, x)
```

The output has two 2×2 matrices.

```
> func2 <- function(x) c(sin(x), cos(x))
> x <- (0:1)*2*pi; x
[1] 0.000000 6.283185
> jacobian(func2, x)
      [,1] [,2]
[1,]    1    0
[2,]    0    1
[3,]    0    0
[4,]    0    0
```

Now we illustrate the convention mentioned above by a multivariate example, where the Jacobian is not a scalar, but a matrix. If $y' = (y_1\ y_2\ y_3)$ and $x' = (x_1\ x_2)$ the convention for derivative of a vector wrt another vector states that we have the following 3×2 matrix:

$$\frac{\partial y}{\partial x} = \begin{bmatrix} \partial y_1/\partial x_1 & \partial y_1/\partial x_2 \\ \partial y_2/\partial x_1 & \partial y_2/\partial x_2 \\ \partial y_3/\partial x_1 & \partial y_3/\partial x_2 \end{bmatrix}. \tag{14.3}$$

Note that even though both y and x are vectors, ($T \times 1$ and $n \times 1$, respectively) the derivative is a $T \times n$ matrix. Observe that i-th row contains the derivatives of the i-th element of y with respect to the elements of x. The numerator is y and its subscript is fixed (=1 for first row, =2 for second row, and so on) along each row according to the convention. To repeat, when both x and y are vectors, the $\partial y/\partial x$ has as many rows (= T) as are the rows in y and as many columns (= n) as the number of rows in x. If one wants to differentiate a matrix wrt a vector or a vector wrt a matrix or a matrix wrt a matrix, it is possible to vectorize the matrices using the vec operator mentioned earlier in Sec. 11.6.1.

In particular, if y is 1×1 (i.e., a scalar), then the above convention (of using the same numerator subscript along each row) implies that $\partial y/\partial x$ has only one row making it a row vector. Thus, according to the convention, $\partial/\partial x$ makes an $n \times 1$ ROW vector of the same dimension as x. In the context of Taylor series and elsewhere it may be convenient to depart from this convention and let $\partial/\partial x$ be a column vector. However, in the present context, if we wish to represent it as a column vector we may do so by considering its transpose, written as: $\partial y/\partial x'$, or $(\partial y/\partial x)'$.

On the other hand, if x is a scalar $(\partial y/\partial x)$ has as many columns as are rows in x, that is, it has only one column making it a column vector.

So far, y was a vector. Now we assume that it is a matrix. If Y is a $T \times p$ matrix and x is still an $n \times 1$ vector, then $\partial Y/\partial x$ is a $Tp \times n$ matrix with as many rows as are total number of elements $(=Tp)$ in the Y matrix.

$$\frac{\partial Y}{\partial x} = \frac{\partial}{\partial x} vec(Y). \tag{14.4}$$

We now state some useful results involving matrix differentiation. Instead of formal proofs, we indicate some analogies with the traditional calculus and argue that they are plausible. Of course, formal proofs available in mathematical textbooks are always needed to be sure. We also include some further comments to help the reader in remembering the correct forms of the results.

In the usual calculus, if $y = 3x$, then $\partial y/\partial x = 3$. Similarly, in matrix algebra we have the following rule: **Derivative of Ax, a matrix times a vector is simply the matrix A**

Let y be a $T \times 1$ vector, x be an $n \times 1$ vector, and let A be a $T \times n$ matrix which does not depend on x. We have

$$y = Ax, \text{ implies } (\partial y/\partial x) = A. \tag{14.5}$$

Note that the i-th element of y is given by

$y_i = \sum_{k=1}^{n} a_{ik}x_k$. Hence, $\{(\partial y_i)/(\partial x_j)\} = \{a_{ij}\}$.

Exercise 1: If $y = a'x = x'a$, where a and x are $T \times 1$ vectors, show that $\partial y/\partial x$ is a row vector according to our convention, and it equals a'.

Exercise 2: If $S = y'y$, where y is a $T \times 1$ vector, show that $\partial S/\partial \beta = 0'$, where β and 0 are $p \times 1$ vectors, since S is not a function of β at all. This is useful in the context of minimizing the error sum of squares u'u in the regression model of Eq. (12.5), $y = X\beta + u$.

Exercise 3: If $S = y'X\beta$, where y is a $T \times 1$ vector, X is a $T \times p$ matrix, and β is a $p \times 1$ vector, show that $\partial S/\partial \beta = y'X$. This is also used in minimizing $u'u$, the residual sum of squares in regression.

14.2   Chain Rule in Matrix Differentiation

In calculus if $y = f(x)$, and $x = g(\theta)$ then $\partial y/\partial \theta = (\partial y/\partial x)(\partial x/\partial \theta)$. Similarly in matrix algebra we have the following chain rule.

If $y = Ax$, as in Eq. (14.5) above, except that the vector x is now a function of another set of variables, say those contained in the r-element column vector θ, then we have the $T \times r$ matrix:

$$\frac{\partial y}{\partial \theta} = \frac{\partial y}{\partial x}\frac{\partial x}{\partial \theta} = A\frac{\partial x}{\partial \theta} \qquad (14.6)$$

where $\frac{\partial x}{\partial \theta}$ is an $n \times r$ matrix following the convention of Eq. (14.3).

Convention for second order derivatives is that one should 'vec' the matrix of first partials before taking second partials.

Let $y = \psi(x)$, where y is a $T \times 1$ vector and x is an $n \times 1$ vector. Now by the second derivative symbol $(\partial^2 y/\, \partial x\, \partial x')$ we shall mean the following $Tn \times n$ matrix:

$$\frac{\partial^2 y}{\partial x \partial x'} = \frac{\partial}{\partial x}vec[\frac{\partial y}{\partial x}] \qquad (14.7)$$

so that the second order partial is a matrix of dimension $(Tn) \times n$. Operationally one has to first convert the $T \times n$ matrix of first partials illustrated by Eq. (14.3) above into a long $Tn \times 1$ vector, and then compute the second derivatives of each element of the long vector with respect to the n elements of x written along the n columns, giving n columns for each of the Tn rows.

14.2.1   *Chain Rule for Second Order Partials wrt θ*

Let $y = Ax$ be as in Eq. (14.5) above. Then we have the following relation:

$$\frac{\partial^2 y}{\partial \theta \partial \theta'} = (A \otimes I_r)\frac{\partial^2 x}{\partial \theta \partial \theta'}. \qquad (14.8)$$

Exercise: True or False ? $\partial^2 y/(\partial \theta\, \partial \theta\,')$ is of dimension $Tnr \times nr$, when $(A \otimes I_r)$ is of dimension $Tn \times n$ and the matrix $\partial^2 x/(\partial \theta\, \partial \theta\,')$ is of dimension $nr \times r$.

Now we consider the first order partial when both x and A are functions of θ. Let $y = Ax$, where y is $T \times 1$, A is $T \times n$, x is $n \times 1$, and both A and x depend on the r-element vector θ. Then the first order partial is the following $T \times r$ matrix:

$$\frac{\partial y}{\partial \theta} = (x' \otimes I_T)\frac{\partial A}{\partial \theta} + A\frac{\partial x}{\partial \theta}. \qquad (14.9)$$

where $\partial A/\partial \theta$ is $Tn \times r$. Next section considers the differentiation of bilinear and quadratic forms

14.2.2 Hessian Matrices in R

The Hessian matrix $H_{ess} = \{h_{ij}\}$ is an $n \times n$ matrix of second order partials of a function $f(x_1, x_2, \ldots, x_n)$, such that $h_{ij} = \partial^2 f/(\partial x_i \partial x_j)$. We have noted the Jacobian matrices in R package 'numDeriv,' Gilbert (2009), before in snippet 14.3.1. The same package also has a function called 'hessian' to calculate a numerical approximation to the Hessian matrix of a function at a parameter value.

The function 'hessian' in the R package called 'maxLik' extracts the Hessian of the maximum likelihood or 'M-estimator' of a statistical model. The R package 'micEcon,' Henningsen (2010) for micro economists has a function 'translogHessian' to compute the Hessian of the translog production function.

14.2.3 Bordered Hessian for Utility Maximization

Consider an economy with n goods x_i and let $p = \{p_i\} > 0$ denote the column vector of per unit prices. The utility function $U(x_1, \ldots, x_n)$ represents a family's level of satisfaction from consuming the set of goods $x = (x_1, \ldots, x_n)$. If we want to maximize a well-behaved (concave) utility function $U(x)$ subject to the linear budget constraint: $p'x \leq Y$, where Y denotes family income.

Section 10.2 shows that constrained optimization leads to determinants of bordered matrices as in Eq. (10.11). A solution to the utility maximization problem $\hat{x}$ here also involves bordered matrices in Eq. (14.10) below. The solution here yields only a local (not global) maximum where the budget constraint is binding and first order conditions are satisfied.

Here the conditions for a maximum are: $(-1)^k det(H_k) > 0$, involving determinants of several bordered Hessian matrices for $k = 2, 3, \ldots, n$. These bordered Hessians must be defined, Simon and Blume (1994, ch. 22) as:

$$H_k = \begin{bmatrix} 0 & \frac{\partial U}{\partial x_1}(\hat{x}) & \cdots & \frac{\partial U}{\partial x_k}(\hat{x}) \\ \frac{\partial U}{\partial x_1}(\hat{x}) & \frac{\partial^2 U}{\partial x_1^2}(\hat{x}) & \cdots & \frac{\partial^2 U}{\partial x_k \partial x_1}(\hat{x}) \\ \vdots & \vdots & \cdots & \vdots \\ \frac{\partial U}{\partial x_k}(\hat{x}) & \frac{\partial^2 U}{\partial x_1 \partial x_k}(\hat{x}) & \cdots & \frac{\partial^2 U}{\partial x_k^2}(\hat{x}) \end{bmatrix}. \tag{14.10}$$

14.3 Derivatives of Bilinear and Quadratic Forms

A bilinear form involves two vectors and a matrix in the middle, usually as $z'Ax$, where we transpose the first vector (prime) times the matrix and then a column vector. Alternative form is to transpose the whole expression as $x'A'z$.

Let $y = z'Ax$, where the bilinear form, y is a 1×1 scalar, z is $T \times 1$, A is $T \times n$, x is $n \times 1$, and A is independent of z and x. Now the $1 \times T$ vector of derivatives is as follows.

$$\frac{\partial y}{\partial z} = x'A', \tag{14.11}$$

and we have the $1 \times n$ vector

$$\frac{\partial y}{\partial x} = z'A. \tag{14.12}$$

Now we turn to evaluating the first derivative of a quadratic form. Let $y = x'Ax$, where x is $n \times 1$, and A is $n \times n$ square matrix and independent of x. Then the partials become a $1 \times n$ vector

$$\frac{\partial y}{\partial x} = x'(A + A'). \tag{14.13}$$

Exercise 1: If A is a symmetric matrix, show that

$$\partial y / \partial x = 2x'A. \tag{14.14}$$

Exercise 2: Show that

$$\partial(\beta'X'X\beta)/\partial\beta = 2\beta'X'X. \tag{14.15}$$

Exercise 3: In a regression model $y = X\beta + \varepsilon$, minimize the error sum of squares defined by $\varepsilon'\varepsilon$ and show that a necessary condition for a minimum is

$$0 = 2\beta'X'X - 2y'X, \tag{14.16}$$

and solve for $\beta = (X'X)^{-1}X'y$. We had encountered this derivation earlier in Eq. (8.32).

14.4 Second Derivative of a Quadratic Form

Let $y = x'Ax$ be a quadratic form, then

$$\frac{\partial^2 y}{\partial x \partial x'} = A' + A, \tag{14.17}$$

and, for the special case where A is symmetric, we have the relation:

$$\frac{\partial^2 y}{\partial x \partial x'} = 2A. \tag{14.18}$$

Now let us consider the first and second derivatives of a bilinear form with respect to θ. If $y = z'Ax$, where z is $T \times 1$, A is $T \times n$, x is $n \times 1$, and both z and x are functions of the r-element vector θ, while A is independent of θ. Then

$$\frac{\partial y}{\partial \theta} = x'A'\frac{\partial z}{\partial \theta} + z'A\frac{\partial x}{\partial \theta}, \tag{14.19}$$

where $\frac{\partial y}{\partial \theta}$ is $1 \times r$, $\frac{\partial z}{\partial \theta}$ is $T \times r$ and $\frac{\partial x}{\partial \theta}$ is $n \times r$. Now the second derivative is as follows.

$$\frac{\partial^2 y}{\partial \theta \partial \theta'} = [\frac{\partial z}{\partial \theta}]'A\frac{\partial x}{\partial \theta} + [\frac{\partial x}{\partial \theta}]'A'[\frac{\partial z}{\partial \theta}] \tag{14.20}$$
$$+(x'A' \otimes I)\frac{\partial^2 z}{\partial \theta \partial \theta'}$$
$$+(z'A \otimes I)\frac{\partial^2 x}{\partial \theta \partial \theta'}.$$

14.4.1 *Derivatives of a Quadratic Form wrt θ*

Consider the quadratic form $y = x'Ax$, where x is $n \times 1$, A is $n \times n$, and x is a function of the r-element vector θ, while A is independent of θ. Then the first derivative of y is:

$$\frac{\partial y}{\partial \theta} = x'(A' + A)\frac{\partial x}{\partial \theta}, \tag{14.21}$$

and the second derivative is:

$$\frac{\partial^2 y}{\partial \theta \partial \theta'} = (\frac{\partial x}{\partial \theta})'(A' + A(\frac{\partial x}{\partial \theta}) + (x'[A' + A] \otimes I)\frac{\partial^2 x}{\partial \theta \partial \theta'}. \tag{14.22}$$

14.4.2 Derivatives of a Symmetric Quadratic Form wrt θ

Consider the same situation as in the subsection above, but suppose in addition that A is symmetric. Then the first derivative is:

$$\frac{\partial y}{\partial \theta} = 2x'A\frac{\partial x}{\partial \theta}, \qquad (14.23)$$

and the second derivative is:

$$\frac{\partial^2 y}{\partial \theta \partial \theta'} = 2(\frac{\partial x}{\partial \theta})'A(\frac{\partial x}{\partial \theta}) + (2x'A \otimes I)\frac{\partial^2 x}{\partial \theta \partial \theta'}. \qquad (14.24)$$

14.4.3 Derivative of a Bilinear Form wrt the Middle Matrix

Let $y = a'Xb$, be the bilinear form, where a and b are $n \times 1$ vectors of constants and X is an $n \times n$ square matrix in the middle. Then the first partial is an $n \times n$ matrix

$$\frac{\partial y}{\partial X} = ab'. \qquad (14.25)$$

14.4.4 Derivative of a Quadratic Form wrt the Middle Matrix

Let $y = a'Xa$,, be a quadratic form where a is an $n \times 1$ vector of constants and X is an $n \times n$ symmetric square matrix in the middle. Then the first partial wrt X is an $n \times n$ matrix:

$$\frac{\partial y}{\partial X} = 2aa' - \text{diag}(aa'), \qquad (14.26)$$

where diag(.) denotes the diagonal matrix based on the diagonal elements of the indicated matrix expression.

14.5 Differentiation of the Trace of a Matrix

Trace of a matrix A is the sum of diagonal elements. When we define the derivative of the trace wrt A itself, we need to focus on differentiation of those diagonal terms. Admittedly the results in this section are not intuitively plausible and one must rely on mathematical proofs from advanced texts.

Let A be a square matrix of order n. In calculus, derivative of x wrt x itself is 1. By analogy, a matrix theory convention states that the partial derivative of the trace of A wrt A itself is the Identity matrix:

$$\frac{\partial tr(A)}{\partial A} = I_n. \tag{14.27}$$

One can differentiate the trace of a product of two matrices, $tr(AB)$, say, with respect to the elements of A. The operation involved can be interpreted as the "rematricization" of the vector in the following sense. Note that $\partial tr(AB)/\partial vec(A)$, is a vector. (Exercise: what dimension ?). Having evaluated it, we need put the resulting vector in a matrix form expressed by $\partial tr(AB)/\partial A$. Some guidance in this matter is available from remembering that if $B = I_n$, then according to Eq. (14.27) the matrix formulated from the derivatives vector should be the identity matrix.

If the elements of A are functions of the r-element vector θ, then the partial of the trace is given by

$$\frac{\partial tr(A)}{\partial \theta} = \frac{\partial tr(A)}{\partial vec(A)} \frac{\partial vec(A)}{\partial \theta} = vec(I)' \frac{\partial vec(A)}{\partial \theta}. \tag{14.28}$$

14.6 Derivatives of $tr(AB)$, $tr(ABC)$

Differentiating the trace of products of matrices with respect to the elements of one of the matrix factors is a special case of differentiation of two types of situations. The first is a linear form: $a'x$, where a is a vector. Second, we will consider the nonlinear form.

(1) Partials of the Trace of Linear Forms:
Let A be $T \times n$, and X be $n \times T$; then

$$\frac{\partial tr(AX)}{\partial X} = A'. \tag{14.29}$$

Exercise: Verify that

$$\frac{\partial tr(A'B)}{\partial B} = A. \tag{14.30}$$

Exercise: Verify this by first choosing $A = I$, the identity matrix.
If X is a function of the elements of the vector θ, then

$$\frac{\partial tr(AX)}{\partial \theta} = \frac{\partial tr(AX)}{\partial vec(X)} \frac{\partial vec(X)}{\partial \theta} = vec(A')' \frac{\partial vec(X)}{\partial \theta} \qquad (14.31)$$

(2) Partials of Trace of Nonlinear (Bilinear) Forms:

Consider the bilinear form: $z'Ax$, where A a matrix in the middle, and z and x appropriately dimensioned vectors. Similarly, the quadratic form is defined as: $x'Ax$.

Now we are ready to discuss the partial derivative of the trace of nonlinear forms involving a product of two or more matrices. Let A be $T \times n, X$ be $n \times q$, and B be $q \times q$; then we have:

$$\frac{\partial}{\partial X} tr(AXB) = A'B'. \qquad (14.32)$$

If x is a function of the r-element vector θ then

$$\frac{\partial}{\partial \theta} tr(AXB) = vec(A'B)' \frac{\partial X}{\partial \theta}. \qquad (14.33)$$

It is remarkable that partial derivatives of the trace of conformably multiplied *four* matrices wrt one of the matrices have been worked out by matrix theorists. Let A be $T \times n, X$ be $n \times q$, B be $q \times r$, and Z be $r \times q$; then

$$\frac{\partial tr(AXBZ)}{\partial X} = A'Z'B', \text{ and} \qquad (14.34)$$

$$\frac{\partial tr(AXBZ)}{\partial Z} = B'X'A'.$$

Viewing the right hand sides of Eq. (14.34), a way to remember the formulas might be to note that right side skips the wrt matrix and transposes all the remaining matrices. Advanced books in matrix theory go on to compute second-order partials involving linear and nonlinear forms.

If X and Z are functions of the r-element vector θ, then

$$\frac{\partial tr(AXBZ)}{\partial \theta} = vec(A'Z'B')' \frac{\partial vec(X)}{\partial \theta} \qquad (14.35)$$
$$+ vec(B'X'A')' \frac{\partial vec(Z)}{\partial \theta}$$

Let A be $T \times T$, X be $q \times T$, B be $q \times q$; then

$$\frac{\partial tr(AX'BX)}{\partial X} = B'XA' + BXA. \qquad (14.36)$$

Exercise: Verify that

$$\frac{\partial tr(X'AXB)}{\partial X} = AXB + A'XB' \tag{14.37}$$

If X is a function of the r-element vector θ, then

$$\frac{\partial tr(AX'BX)}{\partial y} = vec(X)'[(A' \otimes B) \tag{14.38}$$

$$+(A \otimes B')]\frac{\partial vec(X)}{\partial y}.$$

14.6.1 *Derivative* $tr(A^n)$ *wrt* A *is* nA^{-1}

$$\frac{\partial tr(A^n)}{\partial A} = nA^{-1} \tag{14.39}$$

14.7 Differentiation of Determinants

Let A be a square matrix of order T; then

$$\frac{\partial |A|}{\partial A} = A^*, \tag{14.40}$$

where A^* is the matrix of cofactors of A defined in Eq. (6.5) as $C = \{c_{ij}\}$, where its (i,j)-th element is:

$$c_{ij} = (-1)^{i+j} det(A_{ij}), \tag{14.41}$$

where A_{ij} denotes an $(T-1) \times (T-1)$ matrix obtained by deleting i-th row and j-th column of A.

If the elements of A are functions of the r elements of the vector θ, then

$$\frac{\partial |A|}{\partial \theta} = vec(A^*)'\frac{\partial vec(A)}{\partial \theta}. \tag{14.42}$$

14.7.1 *Derivative of* $log(det\ A)$ *wrt* A *is* $(A^{-1})'$

$$\frac{\partial log(det A)}{dA} = (A^{-1})' \tag{14.43}$$

If A is symmetric,

$$\frac{\partial log(det A)}{dA} = 2A^{-1} - diag(A^{-1}), \tag{14.44}$$

where $a_{ij} = a_{ji}$, and where $diag(W)$ denotes a diagonal matrix created from the selection of only the diagonal terms of W.

14.8   Further Derivative Formulas for Vec and A^{-1}

Derivative of a product of three matrices with respect to the vec operator applied to the middle matrix involves a Kronecker product of the transpose of the last and the first matrix.

$$\frac{\partial vec(AXB)}{\partial vecX} = B' \otimes A \qquad (14.45)$$

14.8.1   *Derivative of Matrix Inverse wrt Its Elements*

$$\frac{\partial(X^{-1})}{\partial x_{ij}} = -X^{-1}ZX^{-1} \qquad (14.46)$$

where Z is a matrix of mostly zeros except for a 1 in the $(i,j)^{th}$ position. In this formula if $x_{ij} = x_{ji}$ making it a symmetric matrix, the same formula holds except that the matrix Z has one more 1 in the symmetric $(j,i)^{th}$ position. Why the minus on the right hand side of Eq. (14.46)? When the matrix X is 1×1 or scalar we know from elementary calculus that the derivative of $(1/X)$ is $-1/X^2$. The above formula is a generalization of this result containing the minus.

14.9   Optimization in Portfolio Choice Problem

This section considers Markowitz (1959) celebrated solution to the portfolio choice problem discussed in several Finance textbooks. Our discussion of R in this context is based on Vinod (2008b). The solution is very general, in the sense that available capital funds could be in any plausible amount and any currency.

Assume that there are n possible assets (e.g., stocks and bonds) available for investing. Using data on past returns, investor must allocate available resources to each of these assets. For achieving generality, the allocation problem is formulated as a choice of (positive) weights: $(w_1, w_2, \ldots, w_n)$ so that the weights add up to unity. We have already encountered in Chapter 5 similar choice problems. Of course, this is a stochastic (probabilistic not deterministic) optimization problem involving beliefs about uncertain future returns of assets.

Markowitz (1959) assumes that asset returns are Normally distributed, so that means, variances and covariances describe the entire (multivariate

Normal) density, without having to worry about skewness, kurtosis or other higher moments.

For example, consider only two assets B and C. The random future returns might be denoted by random variables B and C, where the (forecasts of) mean returns are: m_b and m_c, forecast variances are: s_b^2 and s_c^2, and forecast covariance of returns is: COV_{bc}. All are assumed to be available. The risk-adjusted returns can be easily computed if $COV_{bc} = 0$ from the two ratios: (m_b/s_b^2) and (m_c/s_c^2).

Extending this method to any number of assets, a general solution to the optimization problem is obtained by simply ordering the risk-adjusted return ratios and choosing the top (few) assets.

Returning to the example to two assets (without loss of generality) if asset B offers the highest ratio, the optimal choice is to invest all available (money) funds in asset B, and nothing in assets C. That is, the optimal choice allocates the funds in proportion to the following weights $w_b = 1$ and $w_c = 0$. The allocation of half the funds in each asset is stated as: $w_b = 0.5$ and $w_c = 0.5$. The popular 'Sharpe ratio' commonly used by investors worldwide relies on the above solution, except that the risk adjusted return is obtained by dividing by the standard deviation rather than the variance of returns.

Now we write the mathematical objective function for stochastically maximizing the risk-adjusted return of the entire portfolio associated with the allocation proportions: w. For our context of matrix algebra, let us write the objective function in a suitable notation involving matrices and vectors. Recall the discussion of mathematical expectation in section Sec. 5.5. Let ι denote a $n \times 1$ vector of ones and m denote an $n \times 1$ vector of (mean) average returns. Now the expected value of the random variable P 'return of the entire portfolio' based on the allocation w is readily defined.

Initially, it is convenient to consider our example having only two assets. The mathematical expectation of portfolio returns is $E(P) = E(w_b B + w_c C) = w_b m_b + w_c m_c$. In matrix notation, this is written as: $w'm$. The variance of the sum of two random variables is:

$$\text{Var}(w_b B + w_c C) = E[w_b(B - m_b) + w_c(C - m_c)]^2 \tag{14.47}$$
$$= w_b^2 s_b^2 + w_c^2 s_c^2 + 2w_b w_c COV_{ab} = w'Sw,$$

where $w = \{w_a, w_b\}$ and S is the 2×2 variance covariance matrix with (s_b^2, s_c^2) along the diagonal and COV_{ab} as the common off-diagonal term. The variance expression in matrix notation is called a quadratic form in this book at several places. Of course, more generally, S is an $n \times n$ matrix, and w has n elements.

Relaxing the unrealistic assumption of zero covariances among all asset returns is accomplished quite simply by using an S matrix with nonzero off-diagonal elements in the quadratic form. The Lagrangian objective function then is to maximize expected returns $E(P)$ subject to some penalty for volatility $\mathrm{Var}(P)$ involving the quadratic form from Eq. (14.47) and the requirement that elements of w be all positive and sum to 1. We write:

$$\max{}_w(L) = w'm - (1/2)\gamma\, w'Sw - \eta(w'\iota - 1) \qquad (14.48)$$

where γ is Lagrangian coefficient, sometimes called the coefficient of risk aversion and η is the other Langrangian, which depends on the size of the initial capital.

The (necessary) first order condition for maximization of Eq. (14.48) is satisfied by the following solution of the maximization problem. It is obtained by differentiating the Eq. (14.48) with respect to w and setting the derivative equal to zero.

By the rules of matrix algebra discussed above, we evaluate the derivatives of three main terms appearing on the right hand side of Eq. (14.48) (upon ignoring the constants) as:

(1) $(\partial w'm/\partial w)$ is m,
(2) $(\partial w'Sw/\partial w)$ is $2Sw$.
(3) $(\partial w'\iota/\partial w)$ is ι.

Now we recall the constants and plug in these results. The solution to the maximization problem is obtained by equating the first order condition to zero. Next, we simply move w to the left hand side because we have set out to determine the optimum allocation of resources. This way, a rather general solution will yield optimum choice of weights denoted by w^*. We have:

$$w^* = \gamma^{-1}S^{-1}(m - \eta), \qquad (14.49)$$

where the solution is not known till the size of the initial capital η and risk aversion parameter γ are somehow known or specified. Although each investor has a unique value for these parameters, the algebraic solution applies to everyone.

We assume that the second order (sufficient) condition for maximization based on the second derivative is satisfied. It is instructive to discuss a version of the solution Eq. (14.49) discussed in Finance texts using a mean-variance diagram (similar to Figure 14.1) of the portfolio theory, with mean return on the vertical axis and its variance on the horizontal axis.

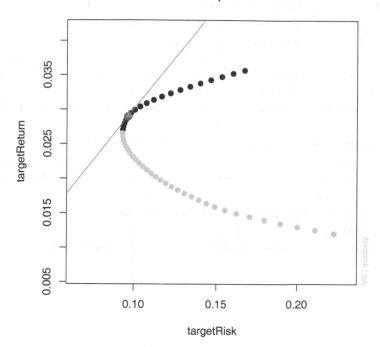

Fig. 14.1 Mean Returns on Vertical Axis and Variance of Returns on Horizontal Axis with Solution on a Concave Frontier

The graphical method makes a distinction between $(n-1)$ risky assets and one asset representing risk-free government bonds. All n assets will have their own mean and variance values, leading to a two-dimensional scatter of n points in this diagram. Since the risk-free return has zero (risk of default) variance, it represents a point on the vertical axis. When n is large, we can lump the scatter of in a few columns for the purpose of discussion. Then, the points representing highest return for each value of the variance will lie along the top part of each column. The assets lying in the lower part of a column are 'dominated,' in the sense that sensible investors will not buy them. Joining the tops of each such column traces a mean-variance 'frontier,' which will be concave (bows out).

The graphical solution is located along the set of all points along the line starting at the point representing risk-free return and ending at the point of tangency of the line with the concave top part of the frontier (See Fig. 14.1). A point at the exact middle of this straight line represents 0.5 weight for both the best risky asset found by the frontier tangency point and on a risk-free government bond.

The Lagrangian γ representing risk aversion will help determine the relative weight on risk-free investment and the asset(s) representing the point of tangency. A risk averse investor will obviously invest a larger share of resources η in risk-free government bonds. This is manifested by choosing a larger weight on that asset.

I have summarized key theoretical insights from portfolio theory. One can use R software to consider more realistic portfolio choices discussed in Vinod (2008b) with additional references to the literature. I also recommend the R package called 'fPortfolio' by Wuertz and Hanf (2010). Since millions of people own at least some (financial) assets, portfolio selection is not a purely theoretical problem. This section has shown that matrix algebra can provide useful insights (if not specific practical solutions) to important real world problems.

Chapter 15

Matrix Results for Statistics

Matrix algebra is very useful for statistical methods as evidenced by Rao (1973). This chapter collects some useful results involving matrix algebra for Statistics with focus on R, which in turn is already focused on Statistics.

15.1 Multivariate Normal Variables

Let us first recall some univariate results. A univariate normal random variable x with mean μ and standard deviation σ usually denoted as $x \sim N(\mu, \sigma^2)$ has the familiar bell-shaped probability density function (pdf) denoted by $f(x)$. It is given by the formula invented in early 19th century as:

$$f(x) = \frac{1}{\sqrt{(2\pi)}\sigma} exp \left[\frac{-(x-\mu)^2}{2\sigma^2} \right]. \tag{15.1}$$

It is also known that the square of a (univariate) Normal variable is a Chi-square variable encountered in Sec. 9.6:

$$[(x-\mu)^2/\sigma^2] \sim \chi_1^2; \qquad x^2 \sim \chi_1^2(\mu^2), \tag{15.2}$$

where the subscript 1 notes that the degrees of freedom is 1 and where μ^2 is the non-centrality parameter of the Chi-square random variable. The central Chi-square variable is denoted by leaving out the non-centrality parameter.

A multivariate variable arises when we consider joint density of two or more variables, Anderson (2003). When $x_1, x_2, \ldots x_n$ are independent and identically distributed (iid) random variables, the joint density is no different from that of individual variables. If we have $x_i \sim N(\mu, \sigma^2)$, with a common mean and variance, then the random variable

$$Y = \Sigma_{i=1}^n [(x_i - \mu)^2/\sigma^2] \sim \chi_n^2, \tag{15.3}$$

is a Chi-square variable with n degrees of freedom, equal to the number of terms in the sum of squares in the definition of Y. The mean and variance of a non-central Chi-square variable are:

$$E[\chi_n^2(\lambda)] = n + \lambda, \quad Var[\chi_1^2(\lambda)] = 2(n + 2\lambda). \tag{15.4}$$

If we relax the assumption of common mean by allowing distinct means μ_i, we have $\lambda = \mu'\mu$ in Eq. (15.4), as the sum of squares of distinct means. The mean and variance of a central Chi-square variable are $n, 2n$, respectively.

Matrix algebra is useful when we consider multivariate situations. Normal density of Eq. (15.1) and all positive Chi-square density are readily evaluated and plotted in R, as shown in the following snippet 15.1.1 for the Normal case. The R function 'dnorm' produces the Normal density for any choice of the mean and standard deviation (sd). The default values are 'mean=0' and 'sd=1,' which need not be specified for the the special case of the "Standard normal density" when $\mu = 0, \sigma = 1$.

A similar R function call 'dchisq(x, df, ncp=0)' where 'ncp' denotes non-centrality parameter allows us to evaluate the density for any specified non-centrality. Using adequate number of evaluations one plots the Chi-square density (not illustrated for brevity). The R function 'rnorm' produces random numbers generated from a Normal parent density, with similarly specified mean and standard deviation. A similar 'rchisq' generates random numbers from Chi-square density.

R program snippet **15.1.1** Normal Density is next.
```
x=seq(-3.5, 3.5, by=0.01)
y=dnorm(x)
plot(x,y,typ="l",ylab="N(0,1) and N(1,2)",main=expression("
Normal densities: "~ mu~" = (0,1); and "~sigma~"=(1,2)"))
y2=dnorm(x, mean=1, sd=2)#non-standard mean and std deviation
lines(x,y2,typ="l", lty=2)
```

The snippet 15.1.1 generates 701 numbers in the closed interval $[-3.5, 3.5]$ by using the 'seq' command for sequence generation in R specifying that the increment be 0.01. The 'y=dnorm(x)' command evaluates Eq. (15.1) for each of the 701 values of x. The option (typ="l") of the plot command is needed to produce a smooth looking *line* plot. Otherwise there will be 701 circles. Note that the choice 'lty=2' produces a dotted line in R. The snippet shows how to get Greek symbols plotted in R figure headings by using the R function 'expression' along with quotes and tildes. R does not allow commas or semicolons, except inside quotes.

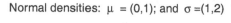

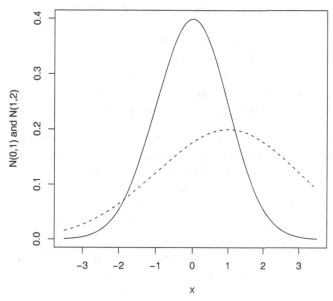

Fig. 15.1   Plot of Eq. (15.1) for $x \in [-3.5, 3.5]$ with 701 evaluations of $y = f(x)$ for the standard Normal case $\mu = 0, \sigma = 1$ as the solid line, and the dotted line for $\mu = 1, \sigma = 2$.

Since the univariate Normal density is completely determined by its mean and variance (square of the standard deviation), it is customary to describe the density of Eq. (15.1) as $x \sim N(\mu, \sigma^2)$, without having to spell out its functional form and using the letter 'N' to represent the Normal density.

Now we consider the bivariate Normal case. Let x and y be two jointly random Normal random variables, with means $E(x)$ and $E(y)$ and variances and covariances arranged in the form of a matrix. We can write

$$\begin{bmatrix} x \\ y \end{bmatrix} \sim N \left[\begin{bmatrix} E(x) \\ E(y) \end{bmatrix}, \begin{bmatrix} V_{xx} & V_{xy} \\ V_{xy} & V_{yy} \end{bmatrix} \right]. \tag{15.5}$$

R package called 'MASS' Venables and Ripley (2002) allows one to simulate from a bivariate Normal parent density and plot it. The following snippet 15.1.2 shows how to use 'kernel density' approximation to simulated data by using the R function 'kde2d' and then plot the simulated density. We will assume that the means and variances are: $\mu = 1, 2, V_{xx} = 1, V_{xy} = 0, V_{yy} = 2$

R program snippet **15.1.2** Bivariate Normal Density is next.
```
library(MASS)
bivn <- mvrnorm(1000, mu = c(0, 1),
Sigma = matrix(c(1, 0, 0, 2), 2))
bivn.kde <- kde2d(bivn[,1], bivn[,2], n = 50)
persp(bivn.kde, phi = 45, theta = 30, shade = 0.2, border = NA,
main="Simulated Bivariate Normal Density")
```

In the snippet 15.1.2, the argument 'Sigma' allows us to specify the V matrix. The function 'kde2d' gets the first and second columns of random numbers in the R object we have named 'bivn' as input coordinates along with 'n=50' for the number of grid points. The function 'persp' draws the perspective or 3D plot as in Fig. 15.1. Its arguments theta, phi are angles defining the viewing direction; theta gives the azimuthal direction and phi the colatitude.

Matrix algebra allows us to generalize the all-important Normal density to the case when x is an $n \times 1$ vector of two or more random variables with a vector of means μ and an $n \times n$ matrix $V = \{v_{ij}, i, j = 1, \ldots n\}$ containing variances along the diagonal (v_{ii}) and covariances along off-diagonal terms ($v_{ij}, i \neq j$) of the symmetric matrix V. The univariate convention involving 'N' is extended to the general case as: $x \sim N(\mu, V)$, with the understanding that x is an n-dimensional (multivariate) normal random variable, with mean vector μ and the $n \times n$ covariance matrix V.

The multivariate Normal density is:

$$f(x) = \frac{1}{(2\pi)^{n/2}|V|^{1/2}} exp[-\frac{1}{2}(x - \mu)'V^{-1}(x - \mu)], \qquad (15.6)$$

where $|V|$ denotes the determinant of V, which is a positive scalar and its square root is well defined.

Our discussion of the bivariate case allows us to consider a partition of the multivariate normal into two components by analogy. Let us partition x into two parts so that x_1, x_2 contain n_1, n_2 components, respectively, satisfying $n = n_1 + n_2$. Now we can write:

$$\begin{bmatrix} x_1 \\ x_2 \end{bmatrix} \sim N \begin{bmatrix} \begin{bmatrix} E(x_1) \\ E(x_2) \end{bmatrix}, \begin{bmatrix} V_{11} & V_{12} \\ V_{12} & V_{22} \end{bmatrix} \end{bmatrix}. \qquad (15.7)$$

Simulated Bivariate Normal Density

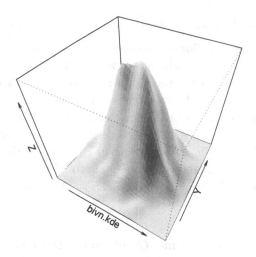

Fig. 15.2   Plot of simulated bivariate Normal when $\mu = 0, 1$ and $V_{xx} = 1, V_{yy} = 2$, $V_{xy} = 0$.

If the n_1 components in x_1 are uncorrelated with the n_2 components in x_2, then $V_{12} = 0 = V_{21}$, are null matrices. Hence, their variance-covariance matrix V is simply block diagonal. The block diagonal property allows us to write the densities of x_1, x_2 independently of each other. For example, the density for the subvector x_1 is:

$$f(x_1) = \frac{1}{(2\pi)^{n_1/2}|V_{11}|^{1/2}} exp\left[\frac{-1}{2}(x - \mu_1)'V_{11}^{-1}(x - \mu_1)\right], \qquad (15.8)$$

where $E(x_1) = \mu_1$. The density for x_2 is written by replacing the subscript 1 by 2 in Eq. (15.8)

A useful property of the Normal density is that a linear transformation of x is also normal. Since linear transformations are the bread and butter of matrix algebra, this property is readily expressed in the vector case as follows.

Consider the following linear transformation of $n \times 1$ vector x into another $n \times 1$ vector y satisfying the relation: $y = Ax + b$, where y and b

are $n \times 1$ vectors and A is a known $n \times n$ matrix involved in the linear transformation. Thus

$$x \sim N(\mu, V) \text{ and } y = Ax + b \Rightarrow y \sim N(b + A\mu, AVA'), \qquad (15.9)$$

where matrix algebra allows us to write the mean and covariance matrix of the transformed variable as indicated in Eq. (15.9). Note that the addition of a vector of constants b to the vector random variable x changes the mean of y but does not affect the variances or covariances at all. This is why the covariance matrix AVA' of y does not contain the vector b. Given that the original V is symmetric, it is easy to verify that the covariance matrix for y is symmetric. This important property can be directly verified, since the transpose of the covariance matrix equals itself: $[(AVA')' = AVA']$ when we apply matrix algebra rules for transpose of a product of two or more matrices.

15.1.1 *Bivariate Normal, Conditional Density and Regression*

Conditional probability of an event E_1 given that another event E_2 has occurred is denoted as $Pr(E_1|E_2)$. Since a density is simply a product of probabilities of individual observations, the conditional density is an extension of conditional probability. Recalling the bivariate Normal density of Eq. (15.5) involving two variables y and x, it is instructive to consider the conditional density of y given x, and relate the expectation of the conditional density to the regression of y on x.

A useful property of Normal density is that the density of y conditional on x is also normal.

$$f(y|x) \sim N\big([E(y) + V_{yx}V_{xx}^{-1}(x - E(x))], (V_{yy} - V_{yx}V_{xx}^{-1}V_{xy})\big), \qquad (15.10)$$

where the expression for the conditional mean involves the variance of x and the covariance V_{xy}. Actually, the conditional mean can be viewed as:

$$E(y|x) = E(y) + V_{yx}V_{xx}^{-1}[x - E(x)] = \bar{y} + b(x - \bar{x}), \qquad (15.11)$$

where b is the regression coefficient (regressing y on x) and $E(x) = \bar{x}, E(y) = \bar{y}$ is the other standard notation for the respective means. The familiar bivariate regression equation of y on x is written as $y = a + bx + \epsilon$, with the error term ϵ, satisfying the property that its mean is zero: $E\epsilon = 0$.

Verify that the intercept of the bivariate regression is estimated as $a = \bar{y} - b\bar{x}$ and the slope is estimated as: $b = V_{yx} V_{xx}^{-1}$.

15.1.2 Score Vector and Fisher Information Matrix

The multivariate density Eq. (15.6) for $t = 1, \ldots T$ observations becomes a likelihood function if interpreted as a function of parameters given the data. The derivative of the log of likelihood function (LL) with respect to the i-th parameter $S_i = \partial LL / \partial \theta_i$ is known as the score vector. The (i, j)-th element of Fisher's information matrix $I_{nfo} = E(S_i S_j)$.

We illustrate this in the context of the familiar regression problem: $y = X\beta + \epsilon, E(\epsilon) = 0. E(\epsilon\epsilon') = \sigma^2 I$, where the density of the t-th observation is Normal:

$$f(y_t | \beta, \sigma) = (\sqrt{2\pi}\sigma)^{-1} exp \left(\frac{-(y_t - X_t\beta)^2}{2\sigma^2} \right), \tag{15.12}$$

where X_t denotes the t-th row of data on all regressors. The log-likelihood function created from the multiplication of densities in Eq. (15.12) over all $t = 1, \ldots, T$ observations becomes:

$$LL(\beta, \sigma | y, X) = -(T/2)log\, 2\pi - (T/2)log\, \sigma^2 - (1/2\sigma^2)[(y - X\beta)'(y - X\beta)], \tag{15.13}$$

upon writing the error sum of squares in matrix notation.

The following score vector is separated into first p components and one component:

$$\partial LL / \partial \beta = (1/\sigma^2) X'(y - X\beta)$$
$$\partial LL / \partial \sigma = (-T/\sigma) + (1/\sigma^3)(y - X\beta)'(y - X\beta) \tag{15.14}$$

The covariance matrix of the scores is the Fisher Information matrix. It is the expectation of the outer product of the score vector with itself. The Information matrix for the two sets of parameters (β, σ) is written as the following 2×2 block matrix:

$$I_{nfo} = \begin{bmatrix} (X'X)/\sigma^2 & 0 \\ 0 & 2/\sigma^2 \end{bmatrix}. \tag{15.15}$$

where we evaluate the squares of the scores and simplify by using the properties of the Normal density.

15.2 Moments of Quadratic Forms in Normals

Let $x = (x_1, x_2, \ldots, x_p)'$ denote a $p \times 1$ column vector of multivariate normal variables with mean vector μ and a symmetric covariance matrix V, having elements V_{ij}, where $i, j = 1, \ldots, p$. That is, $x \sim N(\mu, V)$. One can simplify the derivations and expressions in this section by assuming that the random variable x is measured from its mean, so that $x \sim N(0, V)$. There is no serious loss of generality in this operation.

Let $A = \{a_{ij}\}$ and $B = \{b_{ij}\}$ be $p \times p$ matrices of constants involved in two quadratic forms. Recall that a quadratic form is a scalar obtained by the following matrix multiplication: $Q_A = (x'Ax)$.

For example, if $p = 2$, we have $Q_A = a_{11}x_1^2 + a_{12}x_1x_2 + a_{21}x_1x_2 + a_{22}x_2^2$. First let us write the expression for the expectation (mean) of this quadratic form as:

$$E(Q_A) = a_{11}V_{11} + a_{12}V_{12} + a_{21}V_{12} + a_{22}V_{22}, \qquad (15.16)$$
$$= trace(AV)$$
$$= trace(VA),$$

where $V_{12} = V_{21}$ by symmetry and the mean vector μ is absent. Note that the mean of the quadratic form involves only the elements in the variance-covariance matrix V and the coefficient matrix A.

Now we check that the trace correctly includes the right terms needed in the expression for $E(Q_A)$ spelled out for the $p = 2$ case in the first row of Eq. (15.16). The top left corner of the matrix multiplication AV has elements $a_{11}V_{11} + a_{12}V_{12}$ obtained from the matrix multiplication of the first row of A with the first column of V. Similarly the bottom right corner element of the matrix multiplication AV contains $a_{21}V_{12} + a_{22}V_{22}$. We need to retain only these two diagonal elements from the top left and bottom right corners from the matrix multiplication AV and ignore all off-diagonal elements.

Recall that the trace of a matrix is the sum of its diagonals, which correctly retains only the diagonal elements we want to retain in the answer and ignore the off-diagonals. This verifies that the mean of a quadratic form $E(Q_A) = trace(AV)$ for the $p = 2$ example. It is easy to extend it to $p > 2$ case and also easy to check that $E(Q_A) = trace(VA)$, based on the symmetry of V.

The first four moments of the quadratic form are more conveniently given in terms of the j-th power of the trace $P^j = [trace(AV)]^j$ and trace of the j-th power $Q^j = trace[(AV)^j]$. Following Magnus (1978) we have these rather elegant and compact formulas:

$$E(x'Ax) = \qquad\qquad tr(AV) = P, \qquad\qquad (15.17)$$
$$E(x'Ax)^2 = \qquad\qquad P^2 + 2Q^2,$$
$$E(x'Ax)^3 = \qquad\qquad P^3 + 6PQ^2 + 8Q^3,$$
$$E(x'Ax)^4 = P^4 + 32PQ^3 + 12(Q^2)^2 + 12(P^2)Q^2 + 48Q^4.$$

Why study the third and fourth moment? The motivation may be given by reviewing some elementary statistics involving these moments of random variables x before turning to quadratic forms in x. Let us assume that the variable is univariate with mean μ and variance $E(x-\mu)^2 = \sigma^2$. If the distribution is not symmetric and positively skewed, then the standard measure of skewness based on the third moment is defined for any random variable (not necessarity normal) as:

$$Skew(x) = E(x-\mu)^3/\sigma^3. \qquad\qquad (15.18)$$

The degree of flattening of the density near the center (and fatness of tails) compared to the normal curve is measured by excess kurtosis defined for any random variable as:

$$Kurt(x) = E(x-\mu)^4/\sigma^4 - 3. \qquad\qquad (15.19)$$

For the normally distributed x, it is well known that the distribution is symmtric implying its skewness is zero and has no excess kurtosis, that is: $Skew(x) = 0 = Kurt(x)$.

The univariate definitions of skewness and kurtosis are extended to the multivariate cases in the literature. The quadratic form $Q_A = x'Ax$ in normally distributed multivariate random variable x using the symmetric matrix A in the middle is not at all normally distributed. Hence its skewness and excess kurtosis are worth computing. Collecting some results from Magnus (1978) and denoting the trace by tr we have:

$$E(Q_A) \quad = E(x'Ax) = tr(AV), \qquad\qquad (15.20)$$
$$Var(Q_A) \quad = E(x'Ax)^2 = 2tr(AV)^2,$$
$$Skew(Q_A) = 2\,[2\,tr(AV)^3]^{1/2}[tr(AV)^2]^{-3/2},$$
$$Kurt(Q_A) = 12\,tr(AV)^4\,[tr(AV)^2]^{-2}.$$

Regarding the covariance of two quadratic forms Magnus (1978) has the following result:

$$cov(Q_A Q_B) = E[(x'Ax)(x'Bx)]$$
$$= 2\,tr(AVBV) + tr(AV)tr(BV). \tag{15.21}$$

15.2.1   Independence of Quadratic Forms

Consider two or more quadratic forms where each is a function of the same multivariate Normal variable: $x \sim N(\mu, V)$, where V is a positive definite covariance matrix. Let A, B be n square symmetric matrices. If $AVB = 0$ then we know that $x'Ax$ and $x'Bx$ are independently distributed.

This important result allows partitioning of chi-square random variables in the context of analysis of variance. This result is also important in establishing situations when the ratios of quadratic forms in Normals are distributed as the F random variable, the work horse in significance testing.

15.3   Regression Applications of Quadratic Forms

In usual regression model $y = X\beta + \epsilon$, where y is Normally distributed, errors $\epsilon \sim N(0, \sigma^2 I)$, the hat matrix is $H = X(X'X)^{-1}X'$. The vector of residuals is given by $(I - H)y$. Hence residual sum of squares (RSS) is a quadratic form in Normal random variable y.

$$RSS = y'(I - H)'(I - H)y \tag{15.22}$$

Now RSS is distributed as a Chi-square random variable with degrees of freedom $df = (I - H)'(I - H) = T - p$. If $E(y) = \mu$, then the non-centrality parameter is $\lambda = (1/2)(\mu'(I - H)'(I - H)\mu)$. The matrix algebra formula, $E(y'Ay) = tr(AV)$, implies that the $E(RSS)$ equals $tr(I-H)'(I-H) = tr(I - H) = T - p$, using the property of hat matrix that it is symmetric and idempotent. Recall from Eq. (15.4) that the expected value of a central Chi-square variable equals its degrees of freedom. Anderson (2003) has many advanced results regarding random variables involving quadratic forms, matrix norms, etc.

15.4   Vector Autoregression or VAR Models

Econometricians use vector Autoregressive models as described in Vinod (2008a, p. 228). A VAR model consists of a vector of k endogenous variables, $y_t = (y_{1t}, \ldots, y_{kt})$. Let us define our VAR($p$)-process as:

$$y_t = A_1 y_{t-1} + \cdots + A_p y_{t-p} + a_t, \qquad (15.23)$$

with p vectors of lagged values and p matrices A_i of dimensions $k \times k$ for $i = 1, \ldots, p$, and where $a_t \sim \text{WN}(0, \Sigma)$, a k-dimensional white noise process with Σ denoting a time invariant positive definite covariance matrix.

Stacking of vectors and matrices is useful in the context of VAR models. For example, we stack the dependent variable y_t into $kp \times 1$ vector Y_t with k elements for each of the p lags: $Y_t = [y_t, y_{t-1}, \ldots, y_{t-p+1}]'$. After stacking adequate number of zero vectors, a $VAR(p)$ model of order p becomes equivalent to the following simpler $VAR(1)$ model for theoretical purposes:

$$Y_t = A Y_{t-1} + \epsilon_t, \qquad (15.24)$$

with a large $kp \times kp$ (partitioned) matrix A having the first 'row' containing all the coefficient matrices A_1 to A_p in Eq. (15.23). For example, if $p = 3$ we have the following special case of Eq. (15.24):

$$\begin{pmatrix} y_t \\ y_{t-1} \\ y_{t-2} \end{pmatrix} = \begin{bmatrix} A_1 & A_2 & A_3 \\ I & 0 & 0 \\ 0 & I & 0 \end{bmatrix} \begin{pmatrix} y_{t-1} \\ y_{t-2} \\ y_{t-3} \end{pmatrix} + \begin{pmatrix} u_t \\ 0 \\ 0 \end{pmatrix}. \qquad (15.25)$$

In general, the bottom part of A has a large $k(p-1) \times k(p-1)$ identity matrix along the first $p - 1$ columns of A and the last column (partition) has all zeros. Since Eq. (15.24) can be considered as a first order difference equation, relatively simple methods for solving it become applicable for solving more complicated situations by using suitable stacking of matrices and vectors with suitable partitions.

15.4.1 *Canonical Correlations*

Hotelling discussed the relations between two sets of variables Y, X by using matrix algebra results related to eigenvalues and eigenvectors in 1930s. For example, $Y\beta = \rho X\alpha + \epsilon$ is a regression model with two or more dependent variables as columns in the matrix Y. Certain eigenvalues are transformed and interpreted as correlation coefficients ρ, and the corresponding eigenvectors are reported as analogous to regression coefficients β, α. The theory of canonical correlations is explained in many multivariate statistics texts and monographs, including Vinod (2008a, sec. 5.2) with R examples. R function 'cancor(Y,X)' readily implements Hotelling's canonical correlation model. These are implemented in R by using the QR and singular value decompositions.

15.5   Taylor Series in Matrix Notation

In calculus, Taylor's theorem for a (joint) function of two variables x and y evaluated at x^0, y^0 when the infinitesimal changes in the two variables are, h and k, respectively, is sometimes described by using the D operator notation as follows:

$$f(x,y) = f(x^0, y^0) + Df(x^0, y^0) + \frac{D^2}{2!}f(x^0, y^0) + \ldots + \text{Remainder}, \quad (15.26)$$

where the operator is D. Eq. (15.26) becomes the quadratic approximation for $f(x,y)$ when we ignore all powers of D higher than D^2 and also ignore the remainder term. In particular, the linear operator term is evaluated as:

$$(h\frac{\partial}{\partial x} + k\frac{\partial}{\partial y})f(x^0, y^0) = hf_x(x^0, y^0) + kf_y(x^0, y^0), \quad (15.27)$$

where f_x denotes $\partial f / \partial x$ and f_y denotes $\partial f / \partial y$.

Similarly, the quadratic (2-nd power) term is evaluated as:

$$(h\frac{\partial}{\partial x} + k\frac{\partial}{\partial y})^2 f(x^0, y^0) \quad (15.28)$$
$$= h^2 f_{xx}(x^0, y^0) + 2hk f_{xy}(x^0, y^0) + k^2 f_{yy}(x^0, y^0),$$

where

$$f_{xx} = \partial^2 f / \partial x \partial x, \ f_{xy} = \partial^2 f / \partial x \partial y, \ f_{yy} = \partial^2 f / \partial y \partial y. \quad (15.29)$$

In general,

$$(h\frac{\partial}{\partial x} + k\frac{\partial}{\partial y})^n f(x^0, y^0), \quad (15.30)$$

is obtained by evaluating a Binomial expansion of the n-th power.

Assume that the first n derivatives of a function $f(x,y)$ exist in a closed region, and that the $(n+1) - st$ derivative exists in an open region. Taylor's Theorem states that

$$f(x^0 + h, y^0 + k) = f(x^0, y^0) + (h\frac{\partial}{\partial x} + k\frac{\partial}{\partial y})f(x^0, y^0)$$
$$+ \frac{1}{2!}(h\frac{\partial}{\partial x} + k\frac{\partial}{\partial y})^2 f(x^0, y^0) + \ldots$$
$$+ \frac{1}{n!}(h\frac{\partial}{\partial x} + k\frac{\partial}{\partial y})^n f(x^0, y^0) + \quad (15.31)$$
$$\frac{1}{(n+1)!}(h\frac{\partial}{\partial x} + k\frac{\partial}{\partial y})^{n+1} f(x^0 + \alpha h, y^0 + \alpha k),$$

where $0 < \alpha < 1$, $x = x^0 + h$, $y = y^0 + k$. The term involving the $(n+1)-th$ partial is called the remainder term in Eq. (15.26), where the quadratic approximation ignores all terms containing higher than second order partials.

It is useful to write the Taylor series in the summation and matrix notations. Let x be a $p \times 1$ vector with elements x_1, $x_2, \cdots, x_p$. Now $f(x)$ is a function of p variables. The matrix form of Taylor series needs two building blocks: (i) a vector, and (ii) a matrix. The $p \times 1$ vector is $\partial f(x)/\partial x$ with elements $\partial f(x)/\partial x_i$, where $i = 1, 2, \ldots p$. The $p \times p$ matrix is $\partial^2 f(x)/\partial x \partial x'$ with (i,j)-th element $\partial^2 f(x)/\partial x_i \partial x_j$.

Let us first express Taylor's approximation in the summation notation (without using the matrix algebra) as:

$$f(x) = f(x^o) + \Sigma_{i=1}^{p}(x_i - x_i^o)\frac{\partial}{\partial x}f(x^o) \qquad (15.32)$$

$$+\frac{1}{2}\Sigma_{i=1}^{p}\Sigma_{j=1}^{p}(x_i - x_i^o)[\frac{\partial^2}{\partial x_i \partial x_j}f(x^o)](x_j - x_j^o).$$

Finally we are ready to write the Taylor series in a compact matrix notation after converting summations into matrix and vector multiplications and bringing in a quadratic form as:

$$f(x) = f(x^o) + (x - x^o)'\frac{\partial f(x^o)}{\partial x} \qquad (15.33)$$

$$+\frac{1}{2}(x - x^o)'[\frac{\partial^2 f(x^o)}{\partial x \partial x'}](x - x^o).$$

If the second derivative term is to be the remainder term, we replace $\partial^2 f(x^o)$ by $\partial^2 f(\bar{x})$, where $\bar{x} = \alpha x^o + (1 - \alpha)x$, with $0 \leq \alpha \leq 1$.

Applications of Taylor series are numerous in all branches of sciences including Statistics. For example, Vinod (2008a, p. 200) discusses links between Taylor series and expected utility theory and four moments of probability distribution along with four derivatives of the utility function. Vinod (2008a, p. 382) explains how the delta method for variance of a nonlinear function of a random variable is based on a linear approximation using the Taylor theorem.

Chapter 16

Generalized Inverse and Patterned Matrices

In the matrix algebra discussed so far we have seen that a matrix inverse cannot be computed unless the matrix is square. Here we consider inverting rectangular matrices.

16.1 Defining Generalized Inverse

Before we discuss generalized inverse let us review some properties of the usual inverse. The usual inverse of a square $n \times n$ matrix A satisfies following four properties: (i) $AA^{-1} = I$, (ii) $AA^{-1} - A^{-1}A = 0$, (iii) $AA^{-1}A = A$, and (iv) $A^{-1} = A^{-1}AA^{-1}$. These are indicated in the snippet of the output as Prop.1 to Prop.4.

```
# R program snippet 16.1.1 is next.
set.seed(932); A=matrix(sample(1:9),3,3)
Ainv=solve(A)
A;Ainv
round(A%*%Ainv,7) #is symmetric (Prop.1)
round(A%*%Ainv-t(A%*%Ainv),7) #should be zero
round(Ainv%*%A,7)#is symmetric (Prop.2)
round(Ainv%*%A-t(Ainv%*%A),7) #should be zero
A%*%Ainv%*%A #should be A (Prop.3)
round(A-A%*%Ainv%*%A,7) #should be zero
Ainv%*%A%*%Ainv #should be Ainv Prop.4
round(A-A%*%Ainv%*%A,7) #should be zero
```

```
> A;Ainv
     [,1] [,2] [,3]
```

```
[1,]    3    7    6
[2,]    2    1    4
[3,]    9    8    5
              [,1]         [,2]              [,3]
[1,] -0.18881119  0.09090909  1.538462e-01
[2,]  0.18181818 -0.27272727 -1.059599e-17
[3,]  0.04895105  0.27272727 -7.692308e-02
> round(A%*%Ainv,7) #is symmetric (Prop.1)
     [,1] [,2] [,3]
[1,]    1    0    0
[2,]    0    1    0
[3,]    0    0    1
> round(A%*%Ainv-t(A%*%Ainv),7) #should be zero
     [,1] [,2] [,3]
[1,]    0    0    0
[2,]    0    0    0
[3,]    0    0    0
> round(Ainv%*%A,7)#is symmetric (Prop.2)
     [,1] [,2] [,3]
[1,]    1    0    0
[2,]    0    1    0
[3,]    0    0    1
> round(Ainv%*%A-t(Ainv%*%A),7) #should be zero
     [,1] [,2] [,3]
[1,]    0    0    0
[2,]    0    0    0
[3,]    0    0    0
> A%*%Ainv%*%A #should be A (Prop.3)
     [,1] [,2] [,3]
[1,]    3    7    6
[2,]    2    1    4
[3,]    9    8    5
> round(A-A%*%Ainv%*%A,7) #should be zero
     [,1] [,2] [,3]
[1,]    0    0    0
[2,]    0    0    0
[3,]    0    0    0
> Ainv%*%A%*%Ainv #should be Ainv Prop.4
              [,1]         [,2]              [,3]
```

```
[1,] -0.18881119  0.09090909   1.538462e-01
[2,]  0.18181818 -0.27272727  -2.119198e-17
[3,]  0.04895105  0.27272727  -7.692308e-02
> round(A-A%*%Ainv%*%A,7) #should be zero
      [,1] [,2] [,3]
[1,]   0    0    0
[2,]   0    0    0
[3,]   0    0    0
```

The above output of 16.1.1 verifies the properties Prop.1 to Prop.4 numerically.

Let us begin with a rectangular $m \times n$ matrix A and pretend that its generalized inverse or g-inverse denoted by A^- of dimension $n \times m$ exists and is well defined. Let us temporarily postpone the issue of numerically finding the g-inverse. Rao and Rao (1998, ch.8) has detailed discussion of the theory behind these inverses. Recalling the four properties of a matrix inverse we require that g-inverse also satisfy the following four properties:

$$AA^- = (AA^-)' \text{ or } (AA^-) \text{ is symmetric} \tag{16.1}$$

$$A^- A = (A^- A)' \text{ or } (A^- A) \text{ is symmetric} \tag{16.2}$$

$$AA^- A \quad = A \tag{16.3}$$

$$A^- AA^- \quad = A^- \tag{16.4}$$

If $A = 0$ is a null matrix, we define its g-inverse denoted by $A^- = 0$ to be the null matrix also. Equation (16.1). The Moore-Penrose generalized inverse (often denoted by superscript $(+)$ instead of $(-)$ used above, satisfies the properties Equation (16.2) and Equation (16.3), with the more general statement that they are Hermitian (instead of symmetric). It also satisfies properties Equation (16.4).

16.2 Properties of Moore-Penrose g-inverse

(1) If in particular the matrix A is nonsingular, then Moore-Penrose g-inverse is simply the usual inverse A^{-1}.
(2) In general, the g-inverse of any $n \times m$ rectangular matrix is unique.
(3) The g-inverse of g-inverse gives back the original matrix.
(4) The g-inverse of kA, where k is a constant is simply $(1/k)$ times the g-inverse.
(5) The rank of matrix does not change after g-inverse is computed.

(6) If A^H denotes the conjugate transpose of A, its g-inverse equals the conjugate transpose of the g-inverse of A, that is, $(A^H)^- = (A^-)^H$.

(7) If unitary matrices Σ, Q exist so that PAQ is well-defined. Then its g-inverse is readily found from the relation: $(PAQ)^- = Q^H A^- P^H$.

(8) If two rectangular matrices A, B have a common rank and their product AB is conformable and well defined, then their g-inverse is a product of g-inverses in reverse order: $(AB)^- = B^- A^-$.

(9) The rank of a matrix times its g-inverse satisfies:

$$rank(AA^-) = rank(A) = tr(AA^-) \qquad (16.5)$$

where 'tr' denotes the trace.

(10) The rank of n dimensional Identity minus a matrix times its g-inverse satisfies:

$$rank(I - AA^-) = tr(I - AA^-) = tr(I) - tr(AA^-) = n - rank(X). \qquad (16.6)$$

16.2.1 *Computation of g-inverse*

The simplest way to calculate the g-inverse of A is to use least squares regression formula:

$$A^- = (A'A)^{-1}A' \qquad (16.7)$$
$$= (A^H A)^{-1}A^H \qquad (16.8)$$

where Eq. (16.7) is used when the entries of A are real numbers. If these entries can be complex numbers, we need conjugate transpose of A denoted by A^H present in Eq. (16.8).

Consider the standard regression model,

$$y = X\beta + \epsilon, \qquad (16.9)$$

where β is a $p \times 1$ vector. Let us pre-multiply both sides of Eq. (16.9) by the g-inverse of X defined as $X^- = (X'X)^{-1}X'$, using the formula Eq. (16.7). Note that such pre-multiplication by X^- after ignoring the error term and using the hat notation to denote the estimate gives: $X^-y = X^- X\hat{\beta}$. Since $X^-X = (X'X)^{-1}X'X = I$, this can written as the solution:

$$\hat{\beta} = X^-y = X^- X\hat{\beta}, \qquad (16.10)$$

that is, the estimate $\hat{\beta}$ is the usual $(X'X)^{-1}X'y$, the least squares estimator minimizing the Euclidean norm $\|y - X\beta\|_2$ when the system of underlying equation is consistent. Even for inconsistent cases, the least squares solution is best approximation in some sense.

If we allow for the possibility that

$$(X'X)^-(X'X) \neq I, \tag{16.11}$$

then the solution more general than Eq. (16.10) is:

$$\hat{\beta} = X^-y + [I - X^-X]h, \tag{16.12}$$

where h is an arbitrary $p \times 1$ vector. This gives rise to non-uniqueness of the solution.

The numerical accuracy of Eq. (16.7) is not known to be very good. More on this later in Chapter 17. It is generally recommended that g-inverse be computed by using the singular value decomposition (svd) encountered in Sec. 7.2 and Sec. 12.2.

A $T \times p$ matrix A of rank r has the singular value decomposition (SVD) $A = UDV'$, where U is $T \times p$. The middle $p \times p$ matrix D is diagonal with r nonzero elements denoted by D_1 and $(p - r)$ zero elements. The third $p \times p$ matrix V is orthogonal, containing the eigenvectors. The generalized inverse using the SVD is given by:

$$A^- = V \begin{bmatrix} D_1^{-1} & E \\ F & G \end{bmatrix} U', \tag{16.13}$$

where completely arbitrary matrices E, F, G are of respective dimensions $r \times (T - r)$, $(p - r) \times r$, and $(p - r) \times (T - r)$.

The snippet 16.2.1.1 uses the singular value decomposition for computation of generalized inverse of a randomly defined 4×3 matrix A with full column rank $r = 3$, so that the arbitrary matrices E, F, G from Eq. (16.13) are absent.

```
# R program snippet 16.2.1.1 is next.
set.seed(913); A=matrix(sample(1:100, 12),4,3);A
A.svd <- svd(A)
A.svd$d
ds <- diag(1/A.svd$d[1:3])
u <- A.svd$u
v <- A.svd$v
us <- as.matrix(u[, 1:3])
vs <- as.matrix(v[, 1:3])
A.ginv <- vs%*%(ds)%*%t(us)
```

A.ginv
A%% A.ginv %*% A*

The svd gives us three matrices u, d, v which are accessed by the dollar symbol followed by these names.

The last line of the above snippet verifies that the g-inverse of A satisfies the property $AA^- A = A$.

```
> set.seed(913); A=matrix(sample(1:100, 12),4,3);A
      [,1] [,2] [,3]
[1,]   94   51   59
[2,]   14   91    6
[3,]   25   37    4
[4,]  100   19    2
> A.svd <- svd(A)
> A.svd$d
[1] 166.52213  81.11362  38.69058
> ds <- diag(1/A.svd$d[1:3])
> u <- A.svd$u
> v <- A.svd$v
> us <- as.matrix(u[, 1:3])
> vs <- as.matrix(v[, 1:3])
> A.ginv <- vs%*%(ds)%*%t(us)
> A.ginv
                [,1]            [,2]           [,3]          [,4]
[1,] -0.0001641169 -0.002431743  0.0009382647  0.01026015
[2,] -0.0012677573  0.010517046  0.0035019413 -0.00115618
[3,]  0.0182968315 -0.005350306 -0.0041460763 -0.01541346
> A%*% A.ginv %*% A
      [,1] [,2] [,3]
[1,]   94   51   59
[2,]   14   91    6
[3,]   25   37    4
[4,]  100   19    2
```

The R package called 'MASS' from Venables and Ripley (2002) has a function 'ginv' for computation of the generalized inverse. Hence the following lines of code will yield exactly the same A^- matrix as in the output of snippet 16.2.1.1 given above: 'library(MASS); ginv(A)'. The computations using svd are recommended by numerical mathematicians.

The generalized inverse is not unique. Hence its calculation involves options. Rao (1973, sec. 4i.4) develops a unified theory of least squares using g-inverses to cover situations where 'normal equations' and covariance matrices have singularities. To the best of my knowledge no software is available for their computation. Rao also describes following generalized inverses, among others.

(1) **Minimum Norm g-inverse**: Let the superscript * denotes a conjugate transpose, which becomes a simple transpose when only real numbers are involved. Given a positive definite (p.d.) matrix N, the norm of a vector may be defined as: $\| a \| = (a^* N a)^{1/2}$. Given a consistent equation $Ax = y$, let the minimum norm g-inverse of the matrix A be denoted by $A^-_{m(N)}$, which solves $Ax = y$, where the subscript clarifies that the answer depends on the choice of the p.d. matrix N. It satisfies two conditions: $A A^-_{m(N)} A = A$ and $(A^-_{m(N)} A)^* N = N A^-_{m(N)} A$.

(2) **Least Squares g-inverse**: Given a positive definite (p.d.) matrix N, and given a possibly inconsistent equation $Ax = y$, the solution $x = A^-_{ls(N)} y$ minimizes $(y - Ax)^* N(y - Ax)$, then two conditions are necessary and sufficient: $A A^-_{ls(N)} A = A$ and $(A A^-_{ls(N)})^* N = N A A^-_{m(N)}$.

16.3 System of Linear Equations and Conditional Inverse

Having defined the g-inverse we are now in a position to formally discuss the notion of 'consistency' of a system of equations in real numbers.

Given an $n \times p$ real matrix A and a $p \times 1$ real vector g, we write a general system of n equations in p real unknowns β as:

$$A\beta = g, \tag{16.14}$$

where $A, \beta, g \in \Re$. This system is said to be 'consistent' if a solution β exists, satisfying Eq. (16.14). If not, the system is 'inconsistent,' except that an 'approximate' solution under some definition of 'approximate' might or might not exist.

If A is a non-singular square matrix, we have already seen that the consistent solution $\beta = A^{-1} g$ exists. For example, in the standard regression problem $A = (X'X)$, $g = X'y$ gives 'normal equations' and the solution represents regression coefficients, which do exist (mentioned throughout the book).

If $r = rank(A) = rank[A; G]$, then the vector space spanned by the columns of A is identical to the vector space spanned by columns of $[A; G]$

and the dimension of the vector space equals the rank r, and a consistent solution to Eq. (16.14) exists. This also means that the g vector resides in the column space of A.

Recall that g-inverse of an $n \times p$ matrix A is $p \times n$ and satisfies the property $AA^-A = A$. Matrix theorists including Graybill (1983) use this property to define a new inverse, called conditional inverse (or c-inverse). Thus, c-inverse is defined as:

$$AA^c A = A \qquad (16.15)$$

The c-inverse for each matrix exists, but it may not be unique. A necessary and sufficient condition for a solution to Eq. (16.14) to exist is that the c-inverse A^c satisfies $AA^c g = g$. Since every c-inverse is also a g-inverse, a consistent solution to Eq. (16.14) exists if and only of $AA^- g = g$.

What if the right hand side is a vector of zeros? A system $A\beta = 0$ is known as a homogeneous system. A solution which says that $\beta = 0$ is obviously 'trivial' or not interesting. However, matrix algebra has established that a non-trivial solution exists if and only if $rank(A) < p$ is deficient. Recall in Eq. (9.1) and Eq. (9.2) in the definition of eigenvectors we needed the matrix $(A - \lambda I)$ to be singular (with deficient rank).

Graybill (1983, sec. 7.3) provides the non-unique solutions as:

$$\hat{\beta} = A^c g + [I - A^c A]h, \qquad (16.16)$$

where h is arbitrary $p \times 1$ vector causing the non-uniqueness. More interesting, every solution can be written in this form. Again, Eq. (16.16) also holds if we replace c-inverse by g-inverse.

16.3.1   *Approximate Solutions to Inconsistent Systems*

In practical work, it can happen that a solution to the system of linear equations in real numbers, $A\beta = g$, does not exist. Then, a nonzero error $e(\beta)$ vector satisfies:

$$A\beta - g = e(\beta). \qquad (16.17)$$

Now an approximate solution can be defined by minimizing the error sum of squares (ESS), or $ESS = e(\beta)'e(\beta)$. The minimizing solution is easy to find, since the choice $h = 0$ in Eq. (16.16) can be shown (and visually seen) to minimize the ESS.

16.3.2 *Restricted Least Squares*

Now we consider Eq. (16.9) with m additional (exact linear) restrictions along the m rows of

$$R\beta = r, \tag{16.18}$$

where R is $m \times p$ matrix of rank m and r is $p \times 1$ vector of known real numbers restricting the vector parameters β. We want the restrictions to be 'consistent' in the sense that

$$rank(R) = m \le p. \tag{16.19}$$

Minimizing error sum of squares subject to these additional restrictions yields the Lagrangian minimand:

$$L = (y - X\beta)'(y - X\beta) + \mu'(r - R\beta), \tag{16.20}$$

$$= y'y - 2\beta'X'y + \beta'X'X\beta + \mu'(r - R\beta), \tag{16.21}$$

where μ is an $m \times 1$ vector of Lagrangian multipliers associated with the m constraints along the m rows of Eq. (16.18) The first order conditions (FOC) for minimization of the Lagrangian minimand L are obtained by differentiating L from Eq. (16.21) with respect to β and setting the derivatives equal to zero. Using the derivative formulas discussed later in Eq. (14.5) and Sec. 14.3, we know that

$$\frac{\partial L}{\partial \beta} = 0 = -2X'y + 2X'Xb_R - R'\mu, \tag{16.22}$$

$$\frac{\partial L}{\partial \beta} = 0 = r - Rb_R, \tag{16.23}$$

where the subscripted b_R denotes the restricted least squares (RLS) solution. The next step is to solve Eq. (16.22) and Eq. (16.22) for b_R and μ. To this end, it is convenient to pre-multiply Eq. (16.22) by $(X'X)^{-1}$ and use the definition of ordinary least squares (OLS) estimator $b = (X'X)^{-1}X'y$, from Eq. (8.33) to write Eq. (16.22) as:

$$-2b + 2b_R - (X'X)^{-1}R'\mu = 0. \tag{16.24}$$

Now we pre-multiply Eq. (16.24) by R and use Eq. (16.23) to help remove the Lagrangian coefficient μ by first evaluating it as:

$$\mu = 2[R(X'X)^{-1}R']^{-1}(r - Rb) = 2W^{-1}(r - Rb), \tag{16.25}$$

provided the inverse of the bracketed term W in Eq. (16.25) exists. A necessary (not sufficient) condition for its existence is that $rank(R) = m \leq p$, already assumed in Eq. (16.19).

Our next task is to substitute Eq. (16.25) in Eq. (16.24), cancel the 2 and solve for b_R by moving it to the left hand side to yield the final formula for restricted least squares (RLS) estimator as:

$$b_R = b - (X'X)^{-1}R'W^{-1}(Rb - r). \tag{16.26}$$

The above derivations show the power and usefulness of matrix algebra in linear regression models. It is a good exercise to show that error sum of squares for the restricted least squares estimator $ESS(b_R)$ can be written as the error sum of squares for the ordinary least squares estimator $ESS(b)$ plus a quadratic form in residuals $e = y - Xb$:

$$ESS(b_R) = ESS(b) + e'W^{-1}e, \tag{16.27}$$

which is derived in Vinod and Ullah (1981, ch. 3) and elsewhere. It is known that the mean of b_R is not β, implying that RLS is a biased estimator with a variance covariance matrix $V(b_R)$ having a smaller matrix norm than the variance covariance matrix of the OLS estimator $V(b)$. In other words, $V(b) - V(b_R)$ is a non-negative definite (nnd) matrix. That is, the presence of restrictions can reduce the variance but increases the bias. Vinod and Ullah (1981, ch. 3) contains further discussion of properties of RLS, including the case when there are stochastic and inequality restrictions generalizing Eq. (16.19).

16.4 Vandermonde and Fourier Patterned Matrices

Starting with a set of real numbers $\{x_1, x_2, \ldots x_n\} \in \Re$. The Vandermonde (or alternant) matrix is defined as:

$$A = \{a_{ij}\} = x_j^{i-1}. \tag{16.28}$$

For example, if $n = 3$, Vandermonde matrix is

$$A = \begin{bmatrix} 1 & 1 & 1 \\ x_1 & x_2 & x_3 \\ x_1^2 & x_2^2 & x_3^2 \end{bmatrix}. \tag{16.29}$$

Note that both A and its transpose A' are $n \times n$ Vandermonde matrices. All leading principal submatrices of Vandermonde matrices are also Vandermonde. The determinant of Vandermonde is a product Π of all differences

$(x_j - x_i)$, defined over the range $(n \geq j > i \geq 1)$. Alternatively, we can write:

$$det(A) = \Pi_{t=2}^n \Pi_{i=1}^{t-1}(x_t - x_i) \tag{16.30}$$

From the determinant equation it is easy to check that the rank of Vandermonde matrix equals the number of distinct elements x_i. When the elements are distinct, the inverse of the Vandermonde matrix is analytically known in terms of polynomials $P_i(x)$ of degree $(n-1)$ defined as:

$$P_i(x) = (x - x1)(x - x_2)\ldots(x - x_{i-1})(x - x_{i+1})\ldots(x - x_n)$$
$$= \Pi_{t=1, t \neq i}^n(x - x_t) \quad \text{for} \quad i = 1, 2, \ldots n. \tag{16.31}$$

Note that one can define n polynomials $P_i(x)$ for $(i = 1, 2, \ldots n)$ as follows. For example, if $i = 1, n = 3$, $P_1(x) = (x - x_1)(x - x_2) = x^2 - xx_1 - xx_2 + x_1x_2$. The coefficients of various powers of x can be collected as a_{ij}. For example, when $(i = 1)$, we have $P_1(x) = \Sigma_{j=1}^k a_{ij}x^{j-1}$, with $a_{11} = 1, a_{12} = -(x_1 + x_2)$, and $a_{13} = x_1x_2$. The point is that all $P_i(x)$ are (tedious but) analytically known. The ij-th elements of the inverse of Vandermonde matrix A are analytically known to be $(A^{-1})_{ij} = a_{ij}/P_i(x_i)$.

One application of analytically known inverses is that they can be programmed in a computer for large n and matrix inversion routines can be checked for accuracy by using them.

Another application of (extended) $T \times p$ Vandermonde matrices X is as a matrix of regressors in polynomial regressions with polynomials of order $(p - 1)$. The characteristic polynomials of autoregressive processes often involve polynomial regressions. Once a polynomial regression is specified as: $y = X\beta + \epsilon$, the usual theory applies. If there are T observations the $T \times p$ matrix X of regressors is given for a special case when the third order of polynomial has $p = 4$ regressors with $T > p$ as

$$X = \begin{bmatrix} 1 & x_1 & x_1^2 & x_1^3 \\ 1 & x_2 & x_2^2 & x_2^3 \\ \vdots & \vdots & \vdots \\ 1 & x_T & x_T^2 & x_T^3 \end{bmatrix}. \tag{16.32}$$

Norberg (2010) shows that Vandermonde matrices arise naturally in pricing of zero-coupon bonds, and conditions when they are nonsingular are useful in Finance. The R package 'matrixcalc' has a function for creating Vandermonde matrices as illustrated in the following snippet.

R program snippet **16.4.1** is next.

```
library(matrixcalc)
x=c( .1, .2, .3, .4 )
V <- vandermonde.matrix(x, length(x));V
```

The output of snippet 16.4.1 below shows that the third column has x^2 and fourth column has x^3 values as in Eq. (16.32).

```
> x=c( .1, .2, .3, .4 )
> V <- vandermonde.matrix(x, length(x));V
     [,1] [,2] [,3]  [,4]
[1,]    1  0.1 0.01 0.001
[2,]    1  0.2 0.04 0.008
[3,]    1  0.3 0.09 0.027
[4,]    1  0.4 0.16 0.064
```

16.4.1   *Fourier Matrix*

Matrix theorists, including Graybill (1983, sec.8.12) define a special Vandermonde matrix in terms of various powers of the complex number $w = cos2\pi/n - i\,sin2\pi/n$. The **Fourier matrix** F is defined as $1/\sqrt{n}$ times the $n \times n$ Vandermonde matrix. It can be defined as:

$$
F = \frac{1}{\sqrt{n}}
\begin{bmatrix}
1 & 1 & 1 & \cdots & 1 \\
1 & w & w^2 & \cdots & w^{n-1} \\
1 & w^2 & w^4 & \cdots & w^{2n-1} \\
\vdots & \vdots & \vdots & & \vdots \\
1 & w^{n-1} & w^{2n-2} & \cdots & w^{(n-1)(n-1)}
\end{bmatrix}.
\tag{16.33}
$$

In general, the ij-th element of F is $(1/\sqrt{n})w^{(i-1)(j-1)}$. The Fourier matrix is used in spectral analysis of time series. The following properties of F are known.

(1) The Fourier matrix is symmetric: $F = F'$.
(2) Its inverse equals its conjugate; $F^{-1} = \bar{F}$
(3) $F^2 = P$, where P is $n \times n$ permutation matrix defined in Sec. 16.4.2 containing one of the columns of the $n \times n$ identity matrix.
(4) $F^4 = I$.
(5) The eigenvalues of F are from the set $(1, -1, i, -i)$.

16.4.2 Permutation Matrix

If P is an $n \times n$ square **permutation matrix**, some properties of P are worth noting here.

(1) $P'P = PP' = I$, suggesting that P is orthogonal matrix.
(2) Given some matrix A such that $P'AP$ is well defined, the diagonal elements of $P'AP$ are an rearrangement of the diagonal elements of A.
(3) For a positive integer k, P^k is also a permutation matrix.

16.4.3 Reducible matrix

A square matrix A of dimension $n > 2$ is called reducible matrix if there exist some integer r and an $n \times n$ permutation matrix P defined in Sec. 16.4.2 such that:

$$PAP' = \begin{bmatrix} B & C \\ 0 & D \end{bmatrix}, \tag{16.34}$$

where B is $r \times r$, C is $r \times (n-r)$, and D is $(n-r) \times (n-r)$. The definition permits the all-zero submatrix to be either above or below the main diagonal. If A is not reducible, it is called irreducible. If A is square irreducible matrix, then it must have a positive eigenvalue and the associated eigenvector should have all positive elements.

16.4.4 Nonnegative Indecomposable Matrices

A real $n \times n$ matrix A is called positive (nonnegative) if all its entries are positive (nonnegative). The nonnegative indecomposable matrix is of special interest in some applications.

The definition of decomposable matrix is analogous to that of reducible matrix in Eq. (16.34). The only difference is that one can replace the P' there by a separate permutation matrix Q, permitting greater flexibility.

$$PAQ = \begin{bmatrix} B & 0 \\ C & D \end{bmatrix}. \tag{16.35}$$

where B and D are square and the all-zero submatrix is displayed above the main diagonal instead of below as in Eq. (16.34). The definition permits the all-zero submatrix to be either above or below the main diagonal. If a matrix is not decomposable, it is indecomposable.

16.4.5 *Perron-Frobenius Theorem*

The Perron-Frobenius Theorem proves that an indecomposable real square matrix A having all positive entries $\{a_{ij} > 0\}$ has a unique largest real eigenvalue $\lambda_{max}(A)$ and that the corresponding eigenvector x has strictly positive components. The absolute value of all other eigenvalues satisfy the inequality: $|\lambda_i(A)| < |\lambda_{max}(A)|$. The $\lambda_{max}(A)$ also satisfies the inequalities involving matrix elements:

$min\{a_{ij}\} \leq \lambda_{max}(A) \leq max\{a_{ij}\}$.

Economic theorists are interested in this theorem because they are attracted by the non-negativity of elements of A. After all prices and quantities in Economics are nonnegative. Most applications use the input-output coefficient matrix A discussed in Sec. 8.7. Takayama (1985, ch. 4) describes applications of Perron-Frobenius Theorem in mathematical economics.

Sraffa's 1926 model on 'production of commodities by commodities' uses an irreducible square matrix A of inter-industry commodity flows. Now Sraffa's reference ('numeraire') commodity can be interpreted as the eigenvector of A associated with the largest positive eigenvalue. The eigenvector weights on commodity quantities must be positive to make economic sense, as guaranteed by the Perron-Frobenius Theorem.

16.5 Diagonal Band and Toeplitz Matrices

Section 10.7 discusses bi-diagonal matrices. Here we consider bands of nonzero entries on the main diagonal and some additional nonzero diagonals symmetrically above and below the main diagonal. For example, A_5 is 5-dimensional matrix with five nonzero bands while A_4 is 4-dimensional matrix with three nonzero bands

$$A_5 = \begin{bmatrix} 1 & 4 & 6 & 0 & 0 \\ 2 & 3 & 0 & 7 & 0 \\ 5 & 2 & 3 & 4 & 9 \\ 0 & 1 & 7 & 2 & 9 \\ 0 & 0 & 3 & 4 & 6 \end{bmatrix} ; \quad A_4 = \begin{bmatrix} 1 & 3 & 0 & 0 \\ 4 & 2 & 4 & 0 \\ 0 & 5 & 3 & 4 \\ 0 & 0 & 2 & 4 \end{bmatrix}. \tag{16.36}$$

In general, if A is an $n \times n$ matrix with elements $\{a_{ij}\}$ such that $a_{ij} = 0$ if $|i - j| > k$, where $k < n$ is a positive integer, then A is a band matrix with bandwidth of $2k + 1$. The bandwidth of A_5 in Eq. (16.36) is 5 while that of A_4 is 3. Some properties of band matrices are known in matrix algebra.

(1) A band matrix of bandwidth $2k + 1$ is such that its transpose is also a

band matrix with the same bandwidth.

(2) a weighted sum of band matrices has the same bandwidth provided the weights are scalars.

16.5.1 *Toeplitz Matrices*

Band matrices with the additional property that the values along all diagonals are identical are called Toeplitz matrices. These are produced in R by the function 'toeplitz' and illustrated in the following snippet

```
# R program snippet 16.5.1.1 is next.
set.seed(98); x=sample(1:100,4)
toeplitz(x)
```

The randomly chosen numbers are 47, 35, 33 and 100 and keeping them fixed along the diagonals gives the following matrix.

```
> toeplitz(x)
     [,1] [,2] [,3] [,4]
[1,]   47   35   33  100
[2,]   35   47   35   33
[3,]   33   35   47   35
[4,]  100   33   35   47
```

Analytical expressions for eigenvalues of certain Toeplitz matrices are known in terms of cosines, Graybill (1983, Sec. 8.15). For example if we have a tri-diagonal matrix with super diagonal values a_1 main diagonal values a_0 and sub diagonal values a_2, then m-th eigenvalue is known to be $\lambda_m = a_0 + 2\sqrt{(a_1 a_2)}cos(m\pi/(n+1)), m = 1, 2, \dots, n$.

An application of Toeplitz matrices in regression analysis is as follows. Consider the regression modle $y = X\beta + u$. If true unknown regression errors u follow a first-order autoregressive process, AR(1), $u_t = \rho u_{t-1} + \varepsilon_t$, then the error covariance matrix for u is:

$$\mathrm{Cov}(u) = \Omega = \sigma^2(1-\rho^2)^{-1} \begin{bmatrix} 1 & \rho & \rho^2 & \cdots & \rho^{T-1} \\ \rho & 1 & \rho & \cdots & \rho^{T-2} \\ \vdots & \vdots & \vdots & \vdots & \vdots \\ \rho^{T-1} & \rho^{T-2} & \rho^{T-3} & \cdots & 1 \end{bmatrix}.$$

$$(16.37)$$

which is seen to be a Toeplitz matrix, with all main diagonals having common elements. Its square root matrix is the following patterned bi-diagonal matrix:

$$V_a = \begin{bmatrix} (1-\rho^2)^{1/2} & 0 & 0 & \cdots & 0 & 0 \\ -\rho & 1 & 0 & \cdots & 0 & 0 \\ 0 & -\rho & 1 & \cdots & 0 & 0 \\ \vdots & \vdots & \vdots & \vdots & \vdots & \vdots \\ 0 & 0 & 0 & \cdots & -\rho & 1 \end{bmatrix}. \tag{16.38}$$

16.5.2 Circulant Matrices

A special kind of Toeplitz matrix is circulant. Its unique elements are all in the first row or column. For example, a_0, a_1, a_2, a_3, a_4 give rise to the following 5×5 circulant C:

$$C = \begin{bmatrix} a_0 & a_1 & a_2 & a_3 & a_4 \\ a_1 & a_2 & a_3 & a_4 & a_0 \\ a_2 & a_3 & a_4 & a_0 & a_1 \\ a_3 & a_4 & a_0 & a_1 & a_2 \\ a_4 & a_0 & a_1 & a_2 & a_3 \end{bmatrix}. \tag{16.39}$$

It uses a cyclic permutation matrix defined by

$$P = \begin{bmatrix} 0 & 0 & 0 & 0 & 1 \\ 1 & 0 & 0 & 0 & 0 \\ 0 & 1 & 0 & 0 & 0 \\ 0 & 0 & 1 & 0 & 0 \\ 0 & 0 & 0 & 1 & 0 \end{bmatrix}. \tag{16.40}$$

So that the circulant satisfies the equation:

$$C = a_0 I + a_1 P + a_2 P^2 + \ldots + a_{n-1} P^{n-1}, \tag{16.41}$$

where the polynomial in P is involved. We can check this in the following snippet

```
# R program snippet 16.5.2.1 is next.
P=matrix(0,5,5) #initialize P with all zeros
P[1,5]=1; P[2,1]=1; P[3,2]=1; P[4,3]=1; P[5,4]=1
x=1:5;x; P# view x and the P matrix
C=x[1]*diag(5)+x[2]*P+x[3]*P%*%P+
 x[4]*P%*%P%*%P +x[5]*P%*%P%*%P%*%P; C
```

```
> x=1:5;x; P# view x and the P matrix
[1] 1 2 3 4 5
      [,1] [,2] [,3] [,4] [,5]
[1,]    0    0    0    0    1
[2,]    1    0    0    0    0
[3,]    0    1    0    0    0
[4,]    0    0    1    0    0
[5,]    0    0    0    1    0
> C=x[1]*diag(5)+x[2]*P+x[3]*P%*%P+
+   x[4]*P%*%P%*%P +x[5]*P%*%P%*%P%*%P; C
      [,1] [,2] [,3] [,4] [,5]
[1,]    1    5    4    3    2
[2,]    2    1    5    4    3
[3,]    3    2    1    5    4
[4,]    4    3    2    1    5
[5,]    5    4    3    2    1
```

Among the properties of a circulant C of dimension $n \times n$ we note:

(1) If C, D are circulant matrices, then $C + D, CD$ are also circulant.
(2) If $q_j = a_0 + a_1 w_j + a_2 w_j^2 + \ldots + a_{n-1} w_j^{n-1}$, where $w_j = exp(2\pi i j/n)$ are the n-th roots of unity. Now the determinant of C is a product of these polynomials: $det(C) = \Pi_{j=0}^{n-1} q_j$
(3) Circulant matrices form a commutative algebra in the sense that $CD = DC$
(4) Circulant matrices are related to the discrete Fourier transform and a convolution operator (since roots of unity are involved). Equation (16.37) illustrates a symmetric circulant. For a discussion of symmetric and other circulant matrices see Graybill (1983).

16.5.3 Hankel Matrices

Hankel are special patterned matrices illustrated by

$$H_{ank} = \{h_{ij}\} = \begin{bmatrix} a & b & c & d \\ b & c & d & e \\ c & d & e & f \\ d & e & f & g \end{bmatrix} . \tag{16.42}$$

where the skew-diagonal (90 degrees to the main diagonal) has all constant elements. Note that $h_{ij} = h_{i-1, j+1}$ holds. These matrices have been applied in state space modeling of time series, since a sequence of covariance matrices can be placed in Hankel form, Aoki (1987). Aoki and Havenner (1989) explain that Hankel matrix method is optimal in the sense that the norm of the deviation between the underlying Hankel matrix and its approximation is less than the next omitted singular value. They also show matrix algebraic relations between the Hankel, the 'Principal Components' and 'Canonical Correlations' methods by using singular values of certain matrix multiplications of Hankel-type arranged covariances.

Section 16.7.1 below provides a brief introduction to State Space models focusing on its matrix algebra. Here we note that the rank of the Hankel matrix equals the dimension of the state vector. One limitation of the Aoki's state space model is that it assumes that the matrices A, B, C are fixed over time compared to Eq. (16.47) and Eq. (16.48) in the sequel.

It is interesting that a new R package called 'ebdbNet' by Andrea Rau for 'Empirical Bayes Estimation of Dynamic Bayesian Networks' implements a function called 'hankel', which not only constructs a block-Hankel matrix based on data, but performs a singular value decomposition (SVD) on it and returns the number of 'large' singular values using a flexible cutoff criterion.

16.5.4 *Hadamard Matrices*

Hadamard matrices H are matrices of dimension $n \times n$ with entries 1 or -1 only. They satisfy the property: $H'H = nI$. For example, if $n = 2$ and first two rows of H are $(1,1)$ and $(-1,1)$, respectively, it is easy to verify $H'H = nI$ in R with the commands:

```
h1=c(1,1); h2=c(-1,1); H=rbind(h1,h2);H
t(H) %*% H;
```

The following output of the above code shows that $H'H = nI$ holds for $n = 2$:

```
 h1=c(1,1); h2=c(-1,1); H=rbind(h1,h2);H
    [,1] [,2]
h1    1    1
h2   -1    1
 t(H) %*% H;
     [,1] [,2]
[1,]    2    0
```

[2,]   0   2

Experimental designs is an important branch of Statistics. Since $n^{-1/2}H$ is an orthogonal matrix, these matrices are used for creating so-called orthogonal designs, known to be 'optimal' designs in certain sense. We can list some properties of Hadamard matrices from Graybill (1983).

(1) The dimension of a Hadamard matrix must be $n = 1, 2$ or a multiple of 4. If $n = 1$, $H(1) = 1$.
(2) $n^{-1/2}H$ is an orthogonal matrix.
(3) If D_1, D_2 are $n \times n$ diagonal matrices with entries 1 or -1 only. Then D_1H, HD_2, D_1HD_2 are also Hadamard matrices.
(4) Kronecker products of Hadamard matrices are also Hadamard.
(5) The determinant satisfies $det(H) = n^{n/2}$

The R package 'survey,' Lumley (2004), focused on experimental designs in Statistics has a function 'hadamard' which creates matrices with $1, -1$ or 0 as entries. Since it admits zeros as its elements, it is not a Hadamard matrix in a strict sense. However, it is needed in survey sampling work.

16.6   Mathematical Programming and Matrix Algebra

Important optimization tools in applied science are: linear programming, non-linear programming, integer programming and mixed linear and integer programming. Matrix algebra is needed for effective use of these tools. Objective functions are often stated as quadratic forms and constraints are often stated as linear equalities or inequalities expressed in matrix notation.

Vinod (J1969b) uses integer programming for grouping or clustering of n objects or elements into $m < n$ groups. Each object has some 'value' $p_i, (i = 1, \ldots, n)$ We use an $n \times n$ matrix $\{x_{ij}\}$ of integer elements 1 or 0. Now $x_{ij} = 1$, if i-th element belongs to j-th group; and $x_{ij} = 0$, otherwise. The matrix $\{c_{ij}\}$ measures the 'cost' or 'loss' associated with allocating i-th element to j-th group. The objective is to minimize the total cost. A simplest version minimand is: $\Sigma_{i=1}^n \Sigma_{j=1}^n x_{ij} c_{ij}$.

Vinod (J1969b) also has a 'within group sum of squares' (WGSS) as the minimand and h-dimensional p_i leading to more general definitions of $\{c_{ij}\}$.

The optimizing (integer programming) algorithm allocates n_j elements to the j-th group so that the columns sums $n_j = \Sigma_{i=1}^n x_{ij}$ satisfy: $n = \Sigma_{j=1}^n n_j$, which says that each element must be allocated to some group or another.

Each group has exactly one virtual 'leader' and we define $y_j = 1$ if j-th element is a leader of j-th group. Since there are exactly m groups and satisfy $y_j \geq x_{1j}$, $y_j \geq x_{2j}$, $\ldots y_j \geq x_{nj}$, for all $(j = 1, \ldots, n)$ which makes sure that j-th element is indeed a leader before any element is allocated to it. This will involve n^2 constraints. Fortunately it is possible to formulate them as only n constraints as: $ny_j \geq \Sigma_{i=1}^{n} x_{ij}$ for all $(j = 1, \ldots, n)$.

The elements x_{1j} need to satisfy a String Property: $x_{jj} \geq x_{(j+1),j}$, $x_{jj} \geq x_{(j+2),j}, \ldots x_{jj} \geq x_{nj}$. A solution of the integer programming problem will yield the desired grouping or clustering. Compared to hierarchical clustering this is more computer intensive but globally optimal, since it correctly recognizes that the best cluster need not follow a neat hierarchical order.

The R package 'lpSolve,' Berkelaar *et al.* (2010) has useful algorithms for solving all kinds of mathematical programming problems. The glpk (GNU Linear Programming Kit) package, Lee and Luangkesorn (2009), (written in C as a callable library) solves large-scale linear programming (LP), mixed integer programming (MIP), and related problems.

16.7   Control Theory Applications of Matrix Algebra

16.7.1   *Brief Introduction to State Space Models*

We begin with a brief introduction to state space concepts which use matrix algebra, following Graupe (1972) among others. Autoregressions using difference equations allow economists to model dynamic systems as seen in Sec. 15.4. Engineers often use differential equations to model dynamic systems. Given n initial conditions at time t_0 the state of a dynamic system (involving any number of variables) is described by x, an $n \times 1$ state vector. The input vector to the system is u (also called the forcing function) and coefficient matrices are A, B leading to the system of vector differential equations:

$$\dot{x} = Ax + Bu, \qquad\qquad (16.43)$$

where $\dot{x}$ denotes the vector of derivatives of x with respect to time t.

The simpler homogeneous system of differential equations $\dot{x} = Ax$ with known initial state $x(0) = x_0$ has a well known solution: $x = V \, exp(\lambda t)$, where $V = \{v_{ij}\}$ is an $n \times n$ matrix of eigenvectors of A. Let λ_i, where

$i = 1, 2, \ldots, n$ denote the associated eigenvalues of A. The solution writes $x(t)$ as a system of n equations having v_{ij} coefficients multiplying the $n \times 1$ vector $exp(\lambda t)$, with elements $exp(\lambda_i t)$. R. E. Kalman developed concepts of 'controllability' and 'observability' in 1960s.

Using the eigenvalue-eigenvector decomposition, we write the diagonal matrix $\Lambda = V^{-1}AV$ to diagonalize the system so that Eq. (16.43) becomes

$$\dot{x}^* = \Lambda x^* + B^* u, \tag{16.44}$$

where $B^* = V^{-1}B$. According to Gilbert's criterion, a system in the canonical form from Eq. (16.44) is controllable if no row of B^* is zero. This requires the eigenvalues to be distinct. Otherwise, one needs to work with the diagonal blocks of repeated eigenvalues from the Jordan canonical form mentioned earlier in Sec. 10.7.

Kalman used a solution of the state equation Eq. (16.44)

$$x(t_1) = exp[A(t_1 - t_0)]x(t_0) + \int_{t0}^{t} exp[A(t - \tau)]Bu(\tau)d\tau, \tag{16.45}$$

which involves the expansion containing powers of the matrix A:

$$exp(At) = I + At + A^2 t^2/2! + \ldots + A^k t^k/k! \ldots . \tag{16.46}$$

One uses the Cayley-Hamilton theorem mentioned in Sec. 9.2 to simplify and define the following huge matrix $M = [B; AB; A^2 B; \ldots A^{n-1}B]$. Controllability requires that this matrix be invertible. Matrix algebra plays a crucial role in these results.

While the state vector may not be observable, the output vector of the system is $y = Cx$ is generally observable. In a canonical form the observation equation $y = C^* x^*, C^* = CV$ is added to Eq. (16.44). Such a system is said to be observable if no column of C^* is zero. Observability can also be considered in terms of the expansion of Eq. (16.46) leading to a different huge matrix: $L = [C'; A'C'; (A')^2 C'; \ldots; (A')^{n-1}C']$. Observability requires that the rank of L be n. Engineers are careful to use an 'identified' system where observability and controllability criteria are satisfied and further introduce random (Gaussian) noise as described next.

16.7.2 *Linear Quadratic Gaussian Problems*

Linear quadratic Gaussian problems are important in engineering and control theory. The state of true relations is not known exactly and is subject to additive Gaussian (Normally distributed) errors. The state equation and observation equation are respectively given as:

$$x_{t+1} = A_t x_t + B_t u_t + v_t, \qquad\qquad (16.47)$$

$$y_t = C_t x_t + w_t, \qquad\qquad (16.48)$$

where x_t represents the vector of state variables of the system, the vector of control inputs is u_t and y_t is the vector of measured outputs subject to feedback. The additive error terms v_t, w_t are assumed to be Normally distributed (white noise, Gaussian). Note that coefficient matrices have subscript t suggesting that they are time-varying. The covariance matrices of v_t, w_t are V_t, W_t, also time-varying.

If the time horizon is $t = 0$ to $t = T$, the objective function (minimand) is to minimize the expected value of an expression involving many quadratic forms (where all matrices involved in quadratic forms are assumed to be known) in state variables and control variables:

$$J = E\left(x_T' F x_T + \Sigma_{t=0}^{T-1} x_t' Q_t x_t + u_t' R_t u_t \right). \qquad\qquad (16.49)$$

The solution is recursive starting with mean and variance of the initial state: $\hat{x}_0 = E(x_0)$ and $P_0 = E x_0 x_0'$. Note that P_0 is a symmetric matrix.

The symmetric matrices P_t for $t > 0$ are determined by the following matrix Riccati difference equation that runs forward in time:

$$P_{t+1} = A_t \left(P_t - P_t C_t'[C_t P_t C_t' + W_t]^{-1} C_t P_t \right) A_t' + V_t \qquad\qquad (16.50)$$

Now the Kalman gain can be computed as

$$K_t = A_t P_t C_t'(C_t P_t C_t' + W_t)^{-1}. \qquad\qquad (16.51)$$

Since F in Eq. (16.49) is known, we equate it to $S_T = F$. It could be chosen to be the identity matrix. We then determine $S_{T-1}, S_{T-2}, \ldots$ by using the following matrix Ricatti equation that runs backwards in time.

$$S_t = A_t' \left(S_{t+1} - S_{t+1} B_t[B_t' S_{t+1} B_t + R_t]^{-1} B_t' S_{t+1} \right) A_t + Q_t S_T, \quad (16.52)$$

where the equation is said to run backwards in time because we plug in values at $t + 1$ to get values at t. Using these S_{t+1} values we can get the feedback gain matrix (solution of the minimizing Bellman equation) as

$$L_t = -(B_t' S_{t+1} B_t + R_t)^{-1} B_t' S_{t+1} A_t, \qquad\qquad (16.53)$$

where all items on the right hand side are known. Now we can write the solution as

$$\hat{x}_{t+1} = A_t \hat{x}_t + B_t u_t + K_t(y_t - C_t x_t), \tag{16.54}$$

$$u_t = L_t \hat{x}_t. \tag{16.55}$$

The key point is that we have a solution because all right hand sides can be known, if estimated in a sequence.

These LQG (Linear-quadratic Gaussian) control methods are one of the most useful tools in science and engineering which use matrix algebra. Section 16.5.3 mentions state space modeling using Hankel matrices. An R-forge package called 'robKalman' implements state space modeling using robust Kalman filtering. Hamilton (1994, ch. 13) discusses state space modeling for time series using the Kalman filter, often avoiding the assumption of Gaussian (Normal) distribution.

Integral property of a matrix equation: A matrix equation often arises in identification, optimization and stability analysis problems of interest to engineers as in Graupe (1972, p. 256): Let X, A, B, C be square matrices and define a matrix equation:

$$AX + XA' + C = 0 \tag{16.56}$$

where all the eigenvalues of A have a negative part so that e^A does not diverge. Then X has a unique solution given in terms of improper integrals as:

$$X = \int_0^\infty e^{At} C e^{A't} dt. \tag{16.57}$$

16.8 Smoothing Applications of Matrix Algebra

This section discusses an interesting application of matrix derivatives involving Eq. (16.32) of Sec. 16.4 to the smoothing problem. The well known Hodrick-Prescott (HP) filter, Hodrick and Prescott (1997), is used to derive a smooth version of macroeconomic data to focus on underlying business cycles. Similar smoothing illustrated by Figure 16.1 is of practical interest in many disciplines.

Given a data series: $x_1, x_2, \ldots, x_n$, one is often interested in constructing its smooth version: $y_1, y_2, \ldots, y_n$, which should be fairly close to x_t, but removes excess gyrations among adjacent values. One way to measure changes in adjacent values is in terms of the difference operator $\Delta x_t = (1 - L)x_t$, where L is the lag operator. The objective function of the smoothing problem can be written as a Lagrangian minimand having the Lagrange parameter λ as:

$$L = \Sigma_{t=1}^{n}(x_t - y_t)^2 + \lambda \Sigma_{t=m}^{n}(\Delta^m y_t)^2, \tag{16.58}$$

where the term $(x_t - y_t)^2$ enforces closeness of y_t to the original series, and the last term represents the smoothness penalty. The series y_t becomes more and more smooth as $\lambda > 0$ increases.

Let us rewrite the minimand in matrix notation following Phillips (2010). First, we define a *row* vector:

$$d_m = (mC_0, (-1)mC_1, \ldots, (-1)^{m-1}mC_{m-1}, (-1)^m mC_m), \tag{16.59}$$

where mC_k denotes the Binomial coefficients. For example, when the order of differencing in Eq. (16.58) is $m = 3$, we have: $d_3 = (1, -3, 3, 1)$. Recall that Binomial coefficients can readily come from Pascal triangle or matrix mentioned in Sec. 12.3.2 and implemented in R.

Now define an $(n - m) \times n$ matrix:

$$D' = \begin{bmatrix} d_m & 0 & \ldots & 0 \\ 0 & d_m & \ldots & 0 \\ \vdots & \vdots & \ddots & \vdots \\ 0 & 0 & \ldots & d_m \end{bmatrix} \tag{16.60}$$

Finally, we can write the minimand in Eq. (16.58) as:

$$(x - y)'(x - y) + \lambda y' D D' y \tag{16.61}$$

The solution to the minimization problem based on quadratic form derivatives of Sec. 14.3 gives the first order condition: $(I + \lambda D D')y = x$. Hence we write the solution (smoothed series) as:

$$y = (I + \lambda D D')^{-1} x, \tag{16.62}$$

which depends on the specification of the smoothing Lagrangian $\lambda = (100, 1600, 14400)$, respectively, for annual, quarterly and monthly economic data. The HP filter typically apples to the quarterly data and popularized the choice $\lambda = 1600$. The following R function 'smoother' implements Eq. (16.62) by using R functions 'crossprod' and 'diff'. Its default $\lambda = 1600$ belongs to the HP filter with the default differencing order: $m = 3$.

```
smoother <- function(x,lambda=1600, m=3, lag=1){
ide <- diag(length(x)) #define identity matrix
y <- solve(ide+lambda*crossprod(diff(ide,lag=lag,d=m)),x)
```

```
return(y) }
set.seed(3)
x=cumsum(rnorm(100))
y=smoother(x)
matplot(cbind(x,y),typ="l", main="HP filter Smoothing",
xlab="time sequence",
ylab="Original (solid) and smoothed (dashed)")
```

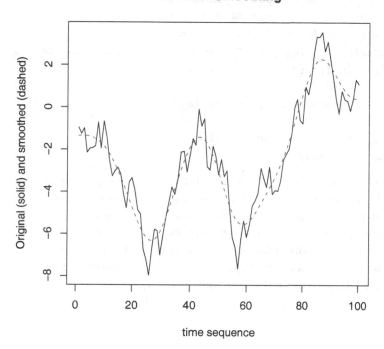

Fig. 16.1 Artificial random walk series x_t (solid line) smoothed by the Hodrick-Prescott filter (dashed line)

Phillips (2010) proves that an approximation to the solution can also be written for the case when $\lambda \to \infty$ as

$$y = [R(R'R)^{-1}R']x, \qquad (16.63)$$

where R is an orthogonal Vandermonde matrix of Eq. (16.32) mentioned above and where one need not specify the λ.

$$R = \begin{bmatrix} 1 & 1 & 1 & \dots & 1 \\ 1 & 2 & 2^2 & \dots & 2^{m-1} \\ \vdots & \vdots & \ddots & \vdots \\ 1 & n & n^2 & \dots & n^{m-1} \end{bmatrix}. \tag{16.64}$$

Using Eq. (16.64), Phillips (2010) argues that as $\lambda \to \infty$, HP filtering does not fully remove the stochastic trend if the data follow a unit root process. We can check this result in R as follows.

```
#assume previous snippet is in memory
y=smoother(x); ou=matrix(NA,9,3);ii=0
PP.test(y) #Phillips-Perron unit root test
colnames(ou)=c("lag","m","  p-value of a unit root test")
for (lag in 1:3){
for (m in 1:3){
ii=ii+1; y2=smoother(x,m=m,lag=lag)
ppy=PP.test(y2)
ou[ii,]=cbind(lag,m, ppy$p.value)}}
rownames(ou)=1:9; ou
```

Note that x_t has a stochastic trend (unit root) by design (being cumulative sum of unit normals) and our HP filter smoothed series y_t should remove it. We test this by using the Phillips-Perron test for the null hypothesis that the series has a unit root against a stationary alternative. The output of the above code (omitted for brevity, showing most p-values exceeding 0.05 for a 5% test) supports Phillips' result based on matrix algebra that the HP filter fails to remove the unit root in the x_t series.

Chapter 17

Numerical Accuracy and QR Decomposition

This chapter focuses on the numerical accuracy issues when digital computers are used and also describes the QR decomposition known for its numerical accuracy. A study of numerical accuracy in matrix computations is a specialized branch of applied mathematics and we can provide only a superficial view here. Computations using paper and pencil are not the same as those using digital computers. There is a natural tendency to simply assume that if an answer is found by using a computer, it must be correct. Unfortunately, numerical analysts have long shown that the assumption is false. The following discussion follows McCullough and Vinod (1999) who wrote "Computers are very precise and can make mistakes with exquisite precision."

17.1 Rounding Numbers

We are familiar with a rounding convention from High School. For example we round 13.5 to 14, but 13.4 to 13. That is, we are taught to always round up when the omitted digit is 5. However, numerical mathematicians have shown that this convention creates a bias. An old solution to this bias is to round to the nearest even number. Unfortunately, it is not easy to program the 'even number rounding'. In the absence of a demand for more accurate arithmetic by the common users, most software programs simply ignore the rounding bias problem.

Apart from this bias, greater care is needed in rounding in general. The Wall Street Journal (8 November 1983, p. 37) reported that the Vancouver Stock Exchange index, (similar to the Dow-Jones Index) began with a nominal value of 1,000.000, recalculated after rounding transactions to four decimal places, the last being truncated so that three decimal places were reported. Intuitively, rounding at the fourth decimal of numbers measured to 10^3 seems innocuous. Yet within a few months the Vancouver index had fallen to 520, without any recession but due to faulty rounding. The correct value should have been 1098.892 according to Toronto Star (29 November 1983).

17.1.1 *Binary Arithmetic and Computer Bits*

While us humans calculate using decimal base-10 arithmetic, computers map all numbers to base 2 (binary) arithmetic (denoted by the subscript '2') where only 0 and 1 are valid numbers. For example $111_2 = 1+2+4 = 7$ $= 2^3\text{-}1$, $1111_2 = 7 + 8 = 2^4 - 1 = 15$, $10000000000_2 = 2^{10} = 1024$.

An eight bit binary number is: $11111111_2 = 2^8 - 1 = 255$, a sixteen bit binary is: $2^{16} - 1 = 65535$. The number of bits stored in computer memory limit the largest binary number possible inside a computer. The largest integer number correctly represented without rounding inside a sixteen-bit computer is only 65535. The numerical inaccuracies arise because base 2 must be combined, as a practical matter, with finite storage and hence finite precision.

Consider a computer "word" with 32 bits, where a floating point binary number is often partitioned as $s \times M \times B^E$, where s is the sign, M is the Mantissa, $B = 2$ is the base and E is the exponent. One bit is for the sign s, eight bits are for E, the signed exponent, which permit $2^7 - 1$ or 127 or -126 as the exponent.

A positive integer Mantissa is a string of ones and zeros, which can be 24-bit long. It can be shown that, McCullough and Vinod (1999, p. 640) real numbers between $(1.2 \times 10^{-38}, 3.4 \times 10^{38})$ alone can be represented in a computer. Numbers outside this range often lead to underflow or overflow errors. It is tempting to avoid the underflow by simply replacing the small number by a zero, but numerical analysis experts do not recommend this.

17.1.2 *Floating Point Arithmetic*

Computers do not have a direct way to represent non-integer numbers involving decimals. This leads to roundoff and cancellation errors in floating point arithmetic. The best representation of the floating decimal number 0.1 in binary format is 0.09999999403953, which does round to 0.1, but the reader is warned that it is only an approximation. By contrast, another decimal $0.5 = 2^{-1}$ needs no approximation, since the binary method readily deals with negative powers of 2. It is intuitively hard to imagine that the numerical accuracy of those fractions which are negative integer powers is excellent, whereas accuracy of nearby numbers is often poor.

The computers are hard wired such that when two floating point numbers are added, the number which is smaller in magnitude is first adjusted by shifting the decimal point till it matches the decimal points of the larger number and then added to efficiently use all available bits. Unfortunately this right shifting of the decimal introduces so-called "roundoff errors."

A particularly pernicious "cancellation error" is introduced when nearly equal numbers are subtracted (the answer which should be close to zero) the error can be of the same magnitude as the answer.

Yet another error called "truncation error" arises because software uses finite term approximations. Consider a long summation of $(1/n)^2$ over a range of $n = 1, 2, \ldots 10000$, by computer software. The result does not necessarily equal the same sum added backwards from $n = 10000, 9999, \ldots, 1$. The experts know that the latter is more accurate, since it starts adding small numbers, where the truncation error is known to be less severe.

17.1.3 *Fibonacci Numbers Using Matrices and Digital Computers*

Fibonacci sequence of integers is named after Leonardo of Pisa, who was known as Fibonacci. His book 'Liber Abaci' published in year 1202 introduced this sequence to Western European mathematics. Fibonacci was interested in population growth of rabbits. The sequence was described in India by Pingala circa 200 BC, whose interest was in the duration of syllables. The sequence is: $x(n) = (0, 1, 1, 2, 3, 5, 8, 13, 21, 34, 55, 89, 144, 233, 377 \ldots)$, where

$$x(n) = x(n-1) + x(n-2), \tag{17.1}$$

that is, n-th number is a sum of previous two numbers, defined for all $n = 3, 4, \ldots$. Similar sequences appear in nature as (i) branching in trees, (ii) arrangement of leaves on a stem, (iii) arrangement of a pine cones, etc. Many fun facts about these numbers are found by searching on Google. One learns that instead of nature trying to use the Fibonacci numbers, they appear as a by-product of deeper physical limitations.

The generation of a long sequence of these numbers is very simple using R as in the following snippet

```
# R program snippet 17.1.3.1 for Fibonacci numbers.
x=rep(NA,100) #initialize x vector of length 100
x[1:2]=c(0,1)#first two Fibonacci numbers
for (i in 3:100){ #begin for loop
x[i]=x[i-1]+x[i-2]} #end for loop
x[1:15]
```

The output reports only the first fifteen numbers for brevity. As we can see, digital computers can accurately handle such sequence of integers. Note that floating point arithmetic of Sec. 17.1.2 is not needed for x.

```
> x[1:15]
[1]   0   1   1   2   3   5   8  13  21  34  55  89 144 233 377
```

Now we employ matrix algebra for generation of these numbers. Since Eq. (17.1) is a two-dimensional second order difference equation:

$$z_{t+1} = \begin{bmatrix} x(n+2) \\ x(n+1) \end{bmatrix} = \begin{bmatrix} 1 & 1 \\ 1 & 0 \end{bmatrix} \begin{bmatrix} x(n+1) \\ x(n) \end{bmatrix} = Az_t, \qquad (17.2)$$

where the reader should note our definitions of vectors z_{t+1}, z_t and the matrix A. Section 8.6 discusses difference equations using matrix algebra.

The matrix A has unit determinant and eigenvalues $(1+w)/2, (1-w)/2$, where $w = \sqrt{5}$. Of course, computation of w needs floating point arithmetic of Sec. 17.1.2. Verify that the sum of eigenvalues equals the trace which is unity. Now the Fibonacci sequence of integers $x(n)$ can be viewed as a solution of the difference equation given by:

$$x(n) = (1/w) \left[(\frac{(1+w)}{2})^{n-1} - (\frac{(1-w)}{2})^{n-1} \right]. \qquad (17.3)$$

In the following snippet we define an R function called Fibo to evaluate Eq. (17.3). It has one input, the value of n. Let us compute the first 100 Fibonacci numbers using the matrix algebra difference equation method.

```
# R snippet 17.1.3.2 Fibonacci numbers by matrix algebra.
#earlier R.snippet should be in memory of R
Fibo=function(n){ #begin Fibo function of n
w=sqrt(5) #define w
out=(1/w)*(0.5+0.5*w)^(n-1) -(1/w)*(0.5-0.5*w)^(n-1)
return(out)}#end Fibo function of n
y=rep(NA,100)#initialize y vector of length 100
y[1:2]=c(0,1)#first two numbers predefined
```

```
for (n in 3:100){
y[n]=Fibo(n)#call the Fibo function above
} #end of for loop
y[1:15]
```

The following output of snippet 17.1.3.2 agrees with the output of snippet 17.1.3.1 for the first fifteen numbers.

```
> y[1:15]
[1]   0   1   1   2   3   5   8  13  21  34  55  89 144 233 377
```

The output of snippet 17.1.3.3 shows that numerical accuracy issues arise for matrix algebra evaluations for large n values. Both earlier snippets must be in the memory of R for this snippet to work. Since we expect that the two methods will give identical numbers we expect $(x - y)$ values to be zero for all n values. The snippet 17.1.3.3 reports a subset of values for various n values indicating that rounding and truncation errors in computer arithmetic for real numbers creeps in.

```
# R program snippet 17.1.3.3 is next.
#R.snippet needs both earlier snippets in memory
n=1:100
cb=cbind(n, x,y, (x-y))#create a 100 by 4 matrix
colnames(cb)[4]=" x-y" #insert name for last column.
nn=seq(10,100,by=10) #define sequence of n values
cb[nn,]#print n, x, y, x-y
xmy=x-y #define x minus y
xmy[70:75]
```

The last column in the following output of snippet 17.1.3.3 should be zero if two versions of Fibonacci numbers are identical in every digit.

```
> cb[nn,]#print n, x, y, x-y
        n            x            y          x-y
[1,]   10 3.400000e+01 3.400000e+01 -1.421085e-14
[2,]   20 4.181000e+03 4.181000e+03 -2.728484e-12
[3,]   30 5.142290e+05 5.142290e+05 -4.656613e-10
[4,]   40 6.324599e+07 6.324599e+07 -7.450581e-08
[5,]   50 7.778742e+09 7.778742e+09 -1.144409e-05
[6,]   60 9.567220e+11 9.567220e+11 -1.831055e-03
[7,]   70 1.176690e+14 1.176690e+14 -2.656250e-01
```

```
[8,]    80 1.447233e+16 1.447233e+16 -4.000000e+01
[9,]    90 1.779979e+18 1.779979e+18 -5.376000e+03
[10,]  100 2.189230e+20 2.189230e+20 -6.881280e+05
> xmy=x-y #define x minus y
> xmy[70:75]
[1]-0.265625 -0.437500 -0.750000 -1.187500 -2.000000 -3.000000
```

As we can see from the output of the snippet 17.1.3.3, the rounding/ truncation errors are creating a discrepancy in accurate x sequence and less
accurate y sequence, the latter using floating point arithmetic. We may
define that a discrepancy $x - y$ in the two numbers creates an erroneous Fibonacci number as soon as $x-y \geq 1$. Starting with $n = 73$, $x-y = -1.1875$
and it becomes as large as -688128 when $n = 100$ for the one hundredth
Fibonacci number. This subsection is a fun illustration of inaccuracies of
floating point arithmetic described in Sec. 17.1.2 and also of solution of
difference equations using matrices.

17.2   Numerically More Reliable Algorithms

In light of the presence of roundoff, truncation and cancellation errors,
careful attention to numerical properties of computer calculations is needed.

Ling (1974) compared algorithms for calculating the variance for a large
data set. Letting Σ denote the sum from $i = 1$ to $i = n$, the numerically
worst formula for variance is the "calculator" formula:

$$V_1 = [1/(n - 1)][\Sigma(x_i)^2 - (1/n)[\Sigma(x_i)]^2]. \qquad (17.4)$$

The standard formula for variance is:

$$V_2 = [1/(n - 1)][\Sigma(x_i - \bar{x})^2], \qquad (17.5)$$

which is more precise, because V_1 squares the observations themselves
rather than their deviations from mean, and in doing so V_1 loses many
more of the smaller bits than V_2 does.

Ling (1974) shows that the following less familiar formula is more accurate. It explicitly calculates $\Sigma(x_i - \bar{x})$, the sum of deviations from the
mean, which can be nonzero only due to cancellation, rounding or truncation errors. The idea is to use the error itself in this calculation to fix the
error in the variance as follows.

$$V_3 = \frac{[\Sigma(x_i - \bar{x})^2 - [1/n]\Sigma(x_i - \bar{x})]}{(n - 1)} \qquad (17.6)$$

Many students find it hard to believe that three algebraically equivalent formulas under paper and pencil algebra methods can be numerically so different, and some even laugh at the idea of adding $[1/n]\Sigma(x_i - \bar{x})$, which should be zero. (Who adds a zero?)

17.3 Gram-Schmidt Orthogonalization

A well-known algorithm, long known to be numerically somewhat unreliable, is the Gram-Schmidt orthogonalization (hereafter "G-SO") algorithm. Recall from Sec. 2.2 that two linearly independent vectors are associated with the usual Cartesian coordinate system of orthogonal axes.

In general, given a specified inner product, every set of linearly independent vectors $\{x_1, x_2, \ldots, x_n\}$ is associated with an orthogonal set of nonzero vectors $\{P_1, P_2, \ldots, P_n\}$. Each vector $P_j, (j = 1, \ldots, n)$ is a linear combination of x_1 through x_{j-1}. Consider the inner product in Eq. (11.2) in our context as $< x_1 \bullet x_1 >_D$ for some D.

(i) The first step at $j = 1$ of G-SO defines $P_1 = (x_1)/[\sqrt{} < x_1 \bullet x_1 >_D]$.

(ii) The next steps of G-SO increase j by 1, where one calculates:
$y_j = x_j - \sum_{i=1}^{j-1} < x_j \bullet P_i >_D P_i$
and sets $P_j = (y_j)/[\sqrt{} < y_j \bullet y_j >_D]$.

(iii) The algorithm stops if $j = n$.

Since the G-SO algorithm is susceptible to roundoff error, a modification called QR algorithm is recommended by experts.

17.4 The QR Modification of Gram-Schmidt

Recall from Sec. 9.5.1 that orthogonal vectors of unit norm are called orthonormal vectors. The QR modification of Gram-Schmidt also transforms a set of linearly independent vectors $\{x_1, x_2, \ldots, x_n\}$ into a set of orthonormal vectors (now denoted with letter Q of the QR algorithm) as: $\{Q_1, Q_2, \ldots, Q_n\}$. The orthonormal Q vectors are such that each Q_j is a linear combination of x_1 through x_{j-1}.

Modified Gram-Schmidt Algorithm: The QR algorithm is iterative and its step for j-th iteration begins by setting $r_{jj} = \|x_j\|_2$, and $Q_j = (1/r_{jj})x_j$.

The step for $k = (j + 1, j + 2, \ldots n)$ we set $r_{jk} = < x_i \bullet Q_j >$.

Also, for $(k = j+1, j+2, \ldots n)$ the algorithm replaces x_k by $x_k - r_{jk}Q_j$. The notation $(\{r_{ij}\}, i < j)$ represents elements of the upper triangular R

matrix of the QR decomposition described next.

17.4.1   *QR Decomposition*

We have discussed the LU decomposition in Sec. 4.10. The QR decomposition is similar but much more popular, because of its reputation as being numerically reliable. It is sometimes viewed as a generalization of Cholesky decomposition discussed earlier in Chapter 11. It applies the modified Gram-Schmidt algorithm described above with steps (i) to (iii) to columns of a matrix A.

The QR decomposition is defined as:

$$A = Q \begin{bmatrix} R \\ 0 \end{bmatrix}, \qquad Q'A = \begin{bmatrix} R \\ 0 \end{bmatrix}, \tag{17.7}$$

where A is $m \times n, m > n$ matrix, Q has orthonormal vectors as columns and R is an upper (right) triangular matrix. In particular, if A is square, then Q is a unitary matrix satisfying $A^H = A^{-1}$, defined in Sec. 11.2.

The QR decomposition of A is immediately available when the columns of A are linearly independent. If not, some of the r_{jj} values are zero. When this happens $Q_j \neq x_j/r_{jj}$ and we must choose Q_j to be any normalized vector orthogonal to $Q_1, Q_2, \ldots, Q_{j-1}$.

The numerical accuracy of QR algorithm might be attributed to our ability to assess when r_{jj} is close to zero. We need to check that the current vector Q_j is orthogonal to $Q_1, Q_2, \ldots, Q_{j-1}$.

17.4.2   *QR Algorithm*

An algorithm for finding the eigenvalues of a real matrix A sequentially constructs matrices $A_j, (j = 1, 2, \ldots)$ by using the QR decompositions of Sec. 17.4.1:

$$A_{j-1} = Q_{j-1}R_{j-1}. \tag{17.8}$$

The next step of the algorithm is to interchange the matrix multiplication terms and define:

$$A_j = R_{j-1}Q_{j-1}. \tag{17.9}$$

Numerical mathematicians have devised ways to accelerate the convergence of the QR algorithm which involve considering a shifted matrix A or using its Hessenberg form.

The QR decomposition plays an important role in many statistical techniques. In particular it can be used to solve the equation $Ax = b$ for given matrix A, and vector b. It is useful for computing regression coefficients and in applying the Newton-Raphson algorithm.

```
# R program snippet 17.4.1 is next.
set.seed(921); X=matrix(sample(1:40,30),10,3)
y=sample(3:13, 10)
XtX=t(X) %*% X; #define XtX= X transpose X
qrx=qr(XtX)#apply qr dcomposition to XtX
Q=qr.Q(qrx);Q #Display Q matrix
#verify that it is orthogonal inverse=transpose
solve(Q)#this is inverse of Q
t(Q) #this is transpose of Q
R=qr.R(qrx); R #Display R matrix
#Note that R is upper triangular
Q %*% R #multiplication Q R
#verify that QR equals the XtX matrix
XtX #this matrix got qr decomposition above.
#apply qr to regression problem
b=solve(t(X) %*% X) %*% (t(X) %*% y);b
qr.solve(X, y, tol = 1e-10)
```

Many regression programs use QR algorithm as default. In snippet 17.4.1 we apply the QR decomposition to $A = X'X$, where X contains regressor data in standard regression problem: $y = X\beta + \epsilon$. We need to use the function 'qr.R' to get the R matrix and 'qr.Q' to get the orthogonal matrix. The snippet verifies that Q is indeed orthogonal by checking whether its transpose equals its inverse, $Q' = Q^{-1}$, numerically. The snippet also checks that the R matrix is upper triangular by visual inspection of the following abridged output.

```
> Q=qr.Q(qrx);Q #Display Q matrix
            [,1]        [,2]        [,3]
[1,] -0.7201992 -0.02614646 -0.6932745
[2,] -0.4810996 -0.70116139  0.5262280
[3,] -0.4998563  0.71252303  0.4923968
> #verify that it is orthogonal inverse=transpose
> solve(Q)#this is inverse of Q
            [,1]        [,2]        [,3]
```

```
[1,]  -0.72019916 -0.4810996 -0.4998563
[2,]  -0.02614646 -0.7011614  0.7125230
[3,]  -0.69327450  0.5262280  0.4923968
> t(Q) #this is transpose of Q
          [,1]        [,2]        [,3]
[1,]  -0.72019916 -0.4810996 -0.4998563
[2,]  -0.02614646 -0.7011614  0.7125230
[3,]  -0.69327450  0.5262280  0.4923968
> R=qr.R(qrx); R #Display R matrix
          [,1]        [,2]        [,3]
[1,]  -10342.97  -6873.9420  -7772.7825
[2,]       0.00   -971.1732   1831.2117
[3,]       0.00      0.0000    548.2268
> #Note that R is upper triangular
> Q %*% R #multiplication Q R
     [,1] [,2] [,3]
[1,] 7449 4976 5170
[2,] 4976 3988 2744
[3,] 5170 2744 5460
> #verify that QR equals the XtX matrix
> XtX #this matrix got qr decomposition above.
     [,1] [,2] [,3]
[1,] 7449 4976 5170
[2,] 4976 3988 2744
[3,] 5170 2744 5460
> #apply qr to regression problem
> b=solve(t(X) %*% X) %*% (t(X) %*% y);b
           [,1]
[1,]  -0.3529716
[2,]   0.5260656
[3,]   0.3405383
> qr.solve(X, y, tol = 1e-10)
[1] -0.3529716  0.5260656  0.3405383
```

QR decomposition or factorization of A can be computed by using a sequence of 'Householder reflections' that successively reduce to zero all 'below diagonal' elements starting with the first column of A, proceeding sequentially to the remaining columns. A Householder reflection (or Householder transformation) starts with an $m \times 1$ vector v and reflects it about some plane by using the formula:

$$H = I - 2\frac{vv'}{\|v\|_2^2}, \tag{17.10}$$

where H is an $m \times m$ matrix.

Let the first reflection yield $H_1\,A$. The second reflection operating on the rows of $H_1\,A$ then yields $H_2\,H_1\,A$, and so on. Maindonald (2010) explains computational devices that avoid explicit calculation of all H matrices in the context of the regression problem: $y_{T\times1} = X_{T\times p}\beta_{P\times1} + \epsilon_{T\times1}$, where matrix dimensions are shown as subscripts.

The R function 'lm' (linear model) to compute estimates of regression coefficients β uses the QR algorithm. It does not even attempt to directly solve the numerically (relatively) unstable 'normal equations,' $X'X\beta = X'y$, as in Eq. (8.33). In fact, the 'lm' function does not even compute the (possibly unstable) $(X'X)$ square matrix used in the snippet 17.4.1 at all. Instead, it begins with the QR decomposition of the rectangular data matrix X is written as in Eq. (17.7). Since Q is orthogonal, its inverse is its transpose and $Q'X$ is also written as in Eq. (17.7). Similar premultiplication by Q' is applied to y also. We write:

$$Q'X = \begin{bmatrix} R \\ 0 \end{bmatrix}, \quad Q'y = \begin{bmatrix} f \\ r \end{bmatrix}, \tag{17.11}$$

where where R is $p \times p$ is upper triangular, 0 is an $T \times p$ matrix of zeros, f is $p \times 1$ vector and r is a $(T - p) \times 1$ vector. Then we write the regression error sum of squares as a matrix norm:

$$\|y - X\beta\|^2 = \|Q'y - Q'X\beta\|^2. \tag{17.12}$$

Now using Eq. (17.11) we can write Eq. (17.12) as $\|f - R\beta\|^2 + \|r\|^2$, which is minimized with respect to β when we choose $\beta = b$, such that $bR = f$. Since R is upper triangular, we have:

$$r_{11}b_1 + r12b_2 + \ldots r_{1p}b_p = f_1$$
$$r22b_2 + \ldots r_{2p}b_p = f_2 \tag{17.13}$$
$$r_{pp}b_p = f_p$$

which is solved from tbe bottom up. First, solve $b_p = f_p/r_{pp}$, then plugging this solution in the previous solution, the algorithm computes b_{p-1}, and so on for all elements of b in a numerically stable fashion.

QR decompositions can also be computed with a series of so-called 'Givens rotations.' Each rotation zeros out an element in the sub-diagonal of the matrix, where zeros are introduced from left to right and from the last row up, forming the R matrix. Then we simply place all the Givens rotations next to each other (concatenate) to form the orthogonal Q matrix.

17.5 Schur Decomposition

We have encountered unitary and Hermitian matrices in Sec. 11.2. Section 10.4 defines 'similar' matrices. Similarity transformations have the property that they preserve important properties of the original matrix including the eigenvalues, while yielding a simpler form. Schur decomposition established that every square matrix is 'similar' to an upper triangular matrix. Since inverting an upper triangular matrix is numerically easy to implement reliably, as seen in Eq. (17.13), it is relevant for numerical accuracy.

For any arbitrary square matrix A there is a unitary matrix U such that

$$U^H AU = U^{-1}AU = T \tag{17.14}$$

$$A = QTQ' \tag{17.15}$$

where T is an upper block-triangular matrix. Matrix A is 'similar' to a block diagonal matrix in Jordan Canonical form discussed in Sec. 10.7.

The R package 'Matrix' defines Schur decomposition for real matrices as in Eq. (17.15). The following snippet generates a 4×4 symmetric Hilbert matrix as our A. Recall from Sec. 12.4 that these matrices are ill-conditioned for moderate to large n values, and hence are useful in checking numerical accuracy of computer algorithms.

```
# R program snippet 17.5.1 Schur Decomposition.
library(Matrix)
A=Hilbert(4);A
schA=Schur(Hilbert(4))
myT=schA@T; myT
myQ=schA@Q; myQ
myQ %*% myT %*% t(myQ)
```

We can check Eq. (17.15) by matrix multiplications indicated there. The following output shows that we get back the original Hilbert matrix.

```
> A=Hilbert(4);A
4 x 4 Matrix of class "dpoMatrix"
          [,1]      [,2]      [,3]      [,4]
[1,] 1.0000000 0.5000000 0.3333333 0.2500000
[2,] 0.5000000 0.3333333 0.2500000 0.2000000
[3,] 0.3333333 0.2500000 0.2000000 0.1666667
[4,] 0.2500000 0.2000000 0.1666667 0.1428571
```

```
> myT=schA@T; myT
4 x 4 diagonal matrix of class "ddiMatrix"
     [,1]      [,2]       [,3]         [,4]
[1,] 1.500214      .          .            .
[2,]       . 0.1691412        .            .
[3,]       .          . 0.006738274        .
[4,]       .          .          . 9.67023e-05
> myQ=schA@Q; myQ
4 x 4 Matrix of class "dgeMatrix"
          [,1]        [,2]        [,3]         [,4]
[1,] 0.7926083  0.5820757 -0.1791863 -0.02919332
[2,] 0.4519231 -0.3705022  0.7419178  0.32871206
[3,] 0.3224164 -0.5095786 -0.1002281 -0.79141115
[4,] 0.2521612 -0.5140483 -0.6382825  0.51455275
> myQ %*% myT %*% t(myQ)
4 x 4 Matrix of class "dgeMatrix"
          [,1]        [,2]        [,3]        [,4]
[1,] 1.0000000 0.5000000 0.3333333 0.2500000
[2,] 0.5000000 0.3333333 0.2500000 0.2000000
[3,] 0.3333333 0.2500000 0.2000000 0.1666667
[4,] 0.2500000 0.2000000 0.1666667 0.1428571
```

Note that the matrix multiplication gives back the original A matrix. The middle matrix T, called 'myT' in the snippet, is seen in the output to be diagonal, a special case of upper block-triangular. This completes a hands-on illustration of the Schur decomposition. For additional practice, the reader can start with an arbitrary square matrix and verify that it can be decomposed as in Eq. (17.15).

Bibliography

Anderson, T. W. (2003). *An Introduction to Multivariate Statistical Analysis*, 3rd edn. (Wiley, New York).

Aoki, M. (1987). *State Space Modeling Time Series* (Springer Verlag, New York).

Aoki, M. and Havenner, A. (1989). A method for approximate representation of vector-valued time series and its relation to two alternatives, *Journal of Econometrics* **42**, pp. 181–199.

Bates, D. and Maechler, M. (2010). *Matrix: Sparse and Dense Matrix Classes and Methods*, URL http://CRAN.R-project.org/package=Matrix, r package version 0.999375-43.

Berkelaar, M. *et al.* (2010). *lpSolve: Interface to Lp_solve v. 5.5 to solve linear/integer programs*, URL http://CRAN.R-project.org/package=lpSolve, r package version 5.6.5.

Bloomfield, P. and Watson, G. S. (1975). The inefficiency of least squares, *Biometrika* **62**, pp. 121–128.

Dimitriadou, E., Hornik, K., Leisch, F., Meyer, D. and Weingessel, A. (2010). *e1071: Misc Functions of the Department of Statistics (e1071), TU Wien*, URL http://CRAN.R-project.org/package=e1071, r package version 1.5-24.

Frank E Harrell Jr and Others. (2010). *Hmisc: Harrell Miscellaneous*, URL http://CRAN.R-project.org/package=Hmisc, r package version 3.8-2.

Gantmacher, F. R. (1959). *The Theory of Matrices*, Vol. I and II (Chelsea Publishing, New York).

Gilbert, P. (2009). *numDeriv: Accurate Numerical Derivatives*, URL http://www.bank-banque-canada.ca/pgilbert, r package version 2009.2-1.

Graupe, D. (1972). *Identification of Systems* (Van Nostrand Reinhold Co., New York).

Graybill, F. A. (1983). *Matrices with Applications in Statistics* (Wadsworth, Belmont, California).

Hamilton, J. D. (1994). *Time Series Analysis* (Princeton University Press).

Henningsen, A. (2010). *micEcon: Microeconomic Analysis and Modelling*, URL http://CRAN.R-project.org/package=micEcon, r package version 0.6-6.

Hodrick, R. and Prescott, E. (1997). Postwar business cycles: an empirical investigation, *Journal of Money Credit and Banking, 29, 1-16.* **29**, pp. 1–16.

Horowitz, K. J. and Planting, M. A. (2006). *Concepts and Methods of the Input-Output Accounts*, 2009th edn. (U.S. Bureau of Economic Analysis of the U.S. Department of Commerce, Washington, DC).

Householder, A. S. (1964). *The Theory of Matrices in Numerical Analysis* (Dover Publications, New York).

James, D. and Hornik, K. (2010). *chron: Chronological Objects which Can Handle Dates and Times*, URL http://CRAN.R-project.org/package=chron, r package version 2.3-35. S original by David James, R port by Kurt Hornik.

Lee, L. and Luangkesorn, L. (2009). *glpk: GNU Linear Programming Kit*, r package version 4.8-0.5.

Leontief, W. (1986). *Input-Output Economics*, 2nd edn. (Oxford University Press, New York).

Ling, R. (1974). Comparison of several algorithms for computing sample means and variances, *Journal of the American Statistical Association* **69**, pp. 859–866.

Lumley, T. (2004). Analysis of complex survey samples, *Journal of Statistical Software* **9**, 1, pp. 1–19.

Magnus, J. (1978). The moments of products of quadratic forms in normal variables, Tech. rep., Open Access publications from Tilburg University, URL http://arno.uvt.nl/show.cgi?fid=29399.

Magnus, J. R. and Neudecker, H. (1999). *Matrix Differential Calculus with Applications in Statistics and Econometrics*, 2nd edn. (John Wiley, New York).

Maindonald, J. (2010). Computations for linear and generalized additive models, URL http://wwwmaths.anu.edu.au/~johnm/r-book/xtras/lm-compute.pdf, unpublished Internet notes.

Marcus, M. and Minc, H. (1964). *A Survey of Matrix Theory and Matrix Inequalities* (Allyn and Bacon, Inc, Boston).

Markowitz, H. (1959). *Portfolio Selection: Efficient Diversification of Investments* (J. Wiley and Sons, New York).

McCullough, B. D. and Vinod, H. D. (1999). The numerical reliability of econometric software, *Journal of Economic Literature* **37**, pp. 633–665.

Norberg, R. (2010). On the vandermonde matrix and its role in mathematical finance, URL http://www.economics.unimelb.edu.au/actwww/html/no75.pdf, department of Statistics, London School of Economics, UK.

Novomestky, F. (2008). *matrixcalc: Collection of functions for matrix differential calculus*, URL http://CRAN.R-project.org/package=matrixcalc, r package version 1.0-1.

Pettofrezzo, A. J. (1978). *Matrices and Transformations*, paperback edn. (Dover, New York, NY).

Phillips, P. C. B. (2010). Two New Zealand pioneer econometricians, *New Zealand Economic Papers* **44**, 1, pp. 1–26.

Rao, C. R. (1973). *Linear Statistical Inference And Its Applications*, 2nd edn. (J. Wiley and Sons, New York).

Rao, C. R. and Rao, M. B. (1998). *Matrix Algebra and Its Applications to Statistics and Econometrics* (World Scientific, Singapore).

Sathe, S. T. and Vinod, H. D. (1974). Bounds on the variance of regression coefficients due to heteroscedastic or autoregressive errors, *Econometrica* **42(2)**, pp. 333–340.

Simon, C. P. and Blume, L. (1994). *Mathematics for Economists* (W. W. Norton, New York).

Soetaert, K. and Herman, P. (2009). *A Practical Guide to Ecological Modelling. Using R as a Simulation Platform* (Springer, New York).

Stewart, R. L., Stone, J. B. and Streitwieser, M. L. (2007). U.s. benchmark input-output accounts, 2002, *Survey Of Current Business* **87**, October, pp. 19–48.

Takayama, A. (1985). *Mathematical Economics*, 2nd edn. (Cambridge University Press, New York).

Venables, W. N. and Ripley, B. D. (2002). *Modern Applied Statistics with S*, 4th edn. (Springer, New York), URL http://www.stats.ox.ac.uk/pub/MASS4, iSBN 0-387-95457-0.

Vinod, H. D. (1968). Econometrics of joint production, *Econometrica* **36**, pp. 322–336.

Vinod, H. D. (1973). A generalization of the durbin–watson statistic for higher order autoregressive process, *Communications in Statistics* **2(2)**, pp. 115–144.

Vinod, H. D. (2008a). *Hands-on Intermediate Econometrics Using R: Templates for Extending Dozens of Practical Examples* (World Scientific, Hackensack, NJ), URL http://www.worldscibooks.com/economics/6895.html, iSBN 10-981-281-885-5.

Vinod, H. D. (2008b). Hands-on optimization using the R-software, *Journal of Research in Management (Optimization)* **1**, 2, pp. 61–65.

Vinod, H. D. (2010). Superior estimation and inference avoiding heteroscedasticity and flawed pivots: R-example of inflation unemployment trade-off, in H. D. Vinod (ed.), *Advances in Social Science Research Using R* (Springer, New York), pp. 39–63.

Vinod, H. D. (J1969b). Integer programming and the theory of grouping, *Journal of the American Statistical Association* **64**, pp. 506–519.

Vinod, H. D. and Ullah, A. (1981). *Recent Advances in Regression Methods* (Marcel Dekker, Inc., New York).

Weintraub, S. H. (2008). *Jordan Canonical Form, Applications to Differential Equations* (Morgan and Claypool publishers, Google books).

Wuertz, D. and core team, R. (2010). *fBasics: Rmetrics - Markets and Basic Statistics*, URL http://CRAN.R-project.org/package=fBasics, r package version 2110.79.

Wuertz, D. and Hanf, M. (eds.) (2010). *Portfolio Optimization with R/Rmetrics* (Rmetrics Association & Finance Online, www.rmetrics.org), r package version 2110.79.

Wuertz, D. *et al.* (2010). *fUtilities: Function Utilities*, URL http://CRAN.R-project.org/package=fUtilities, r package version 2110.78.

Index